NAJET BAKI

Towards a Solar Future

NAJET BAKI

Towards a Solar Future

Exploring the Sun's four energies(passive, photovoltaic, thermal and thermodynamic)

ScienciaScripts

Cover image: www.ingimage.com

This book is a translation from the original published under ISBN 978-3-8416-3734-5.

Publisher:
Sciencia Scripts
is a trademark of
Dodo Books Indian Ocean Ltd. and OmniScriptum S.R.L publishing group

120 High Road, East Finchley, London, N2 9ED, United Kingdom
Str. Armeneasca 28/1, office 1, Chisinau MD-2012, Republic of Moldova, Europe
Printed at: see last page
ISBN: 978-620-8-23299-3

Towards a Solar Future
Exploring the Sun's four energies

" The sun generously offers us its beautiful rays of light every day: let's make the most of it!"

Najet BAKI

BAKI Najet

This book is dedicated to :

To my invaluable parents, whose love and support are beyond what I can express in words. You are the dearest to my heart, and I cherish you more than anything else in the world.

To my wonderful children, who are my greatest inspiration. May you always pursue your dreams with the same determination you've seen in me. This book is for you, so you know that anything is possible.

To my precious family, my sisters and my two brothers, who have always been my cornerstone. Your love, support and encouragement have brightened my path. This book is the fruit of our collective strength and unwavering commitment. May these pages, bathed in the light of our family bond, remind you how dear you are to my heart. With every word, I celebrate our unity and dedication. And never forget that "the sun generously offers us its beautiful, luminous rays every day: let's make the most of them!

With all my love.

Thanks

I would like to express my deep gratitude to all those who contributed to the realization of this book. My most sincere thanks go to my daughter Meriem, whose artistic talent brought this book to life with her magnificent illustrations. Many thanks to my son Mustapha, who contributed his expertise in Python programming to make this project possible. Finally, a warm smile of gratitude to my son Yasser, whose cheerful enthusiasm has been a constant source of inspiration. Your unique contributions made this book a reality, and I'm proud to have you as my children. You are the light that has illuminated this path. Thank you from the bottom of my heart for all you have done.

I would also like to express my deep gratitude to the many books and websites that have been a valuable source of knowledge and inspiration throughout this project. Their rich content has informed my journey and enabled me to create this book. I am grateful for the invaluable resources they have provided.

Purpose of this book:

The sun generously offers us its beautiful rays of light every day: let's make the most of it!

This book aims to explore the four forms of solar energy, including photovoltaic, thermal, thermodynamic and passive solar energy, which is a solution not to be overlooked! It presents the basic principles, how they work, their applications, and their potential for integration into various environments. Through case studies, this book guides the reader through the challenges of optimizing the capture and use of solar energy. It also emphasizes the importance of energy efficiency, sustainability and innovation in the development of solar systems, in order to contribute to a cleaner, more responsible energy future. And at the end of the manuscript, there are ten exercises with solutions.

Table of contents

General introduction :

Solar energy is a highly diversified renewable resource, exploited through a variety of technologies to meet diverse energy needs. Among these technologies, there are four main types of solar energy: passive, active (thermal, thermodynamic and photovoltaic). Passive solar energy exploits architectural design to maximize the direct use of solar light and heat. The second, active thermal, uses thermal collectors to heat fluids for domestic use. The third, active thermodynamics, concentrates sunlight to produce heat and electricity on a large scale. Finally, the fourth, photovoltaics, converts sunlight directly into electricity using photovoltaic cells. Each form of solar energy offers specific advantages to suit local energy needs and environmental conditions, contributing to the transition to more sustainable, environmentally-friendly energy systems and to the reduction of greenhouse gas emissions.

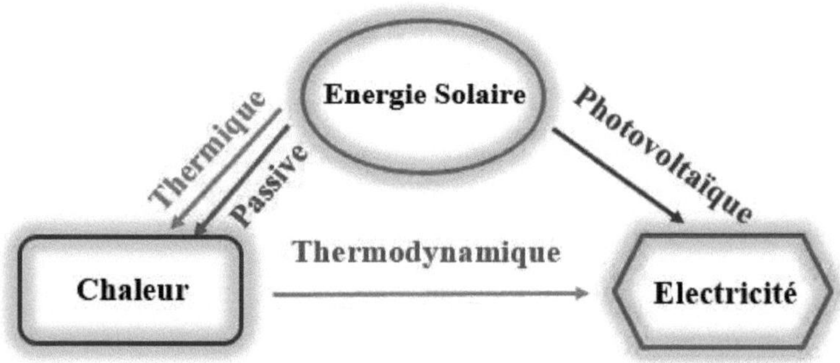

General information on solar energy

1-Introduction :

At a time when the climate crisis is growing and fossil fuel reserves are dwindling, the issue of energy transition is emerging as a global priority. This manuscript explores the fundamentals of renewable energies, not only as a response to environmental challenges, but also as an opportunity to redefine our relationship with energy. Understanding the evolution of these alternative energy sources, as well as their untapped potential, is crucial to building a sustainable energy future, capable of meeting humanity's growing needs while preserving the ecosystems on which we depend.

2-World energy :

The global energy landscape is dominated by fossil fuels, such as oil, coal and natural gas, which provide the majority of the energy used to power industry, transport and households. However, this dependence creates critical challenges, notably in terms of energy security, price volatility and long-term availability. At the same time, renewable energies, although still representing a modest share of the world's energy mix, are experiencing rapid growth thanks to technological innovations and pressure to reduce greenhouse gas emissions.

3-Fossil fuels and environmental impact :

Fossil fuels, while fundamental to industrial development, have a high environmental cost. Their combustion releases massive quantities of carbon dioxide (CO_2) and other pollutants, contributing to global warming, air pollution and ocean acidification. What's more, the extraction of these resources often leads to environmental degradation, such as deforestation, oil spills and the destruction of natural habitats.

These harmful impacts underline the urgency of a transition to cleaner, more sustainable energies.

4- From the oil era to the post-oil era :

With the gradual exhaustion of oil reserves and the recognition of climate risks, the world is heading towards a new energy era, often referred to as "post-oil". This transition is marked by a diversification of energy sources, increased use of renewable energies, and improved energy efficiency. Public policy and technological innovation are playing a crucial role in this transition, promoting the search for alternative solutions to replace oil in key sectors such as transport and power generation.

5-Exploitation of renewable energies :

Renewable energies are energy sources that regenerate naturally on a human timescale, making them virtually inexhaustible compared with fossil fuels. These sources include solar, wind, hydraulic, geothermal and biomass energy. Their exploitation relies on specific technologies: solar panels to capture solar energy, wind turbines to transform wind energy into electricity, dams to harness hydraulic energy, and so on. These technologies make it possible to produce energy with a reduced environmental impact, which is essential for a sustainable energy transition.

5.1-Green energy :

Green energies are renewable energy sources that do not run out with use, aiming to replace fossil fuels and reduce greenhouse gas emissions responsible for climate change. The use of renewable energies dates back to civilizations such as the Greeks and Romans, who used water power to grind grain and wind power to run windmills. However, the modern concept and development of renewable energies is more recent.

Figure 1: Renewable energy.

5.2-The different types of renewable energy :

***Solar** energy: solar energy is captured from the sun's rays and can be converted into electricity using photovoltaic solar panels, or used directly to heat water or air using solar thermal systems.

***Wind** energy: Wind energy is generated by the force of the wind, which turns wind turbines. Wind turbines convert the wind's kinetic energy into electricity.

***Biomass** energy: Biomass energy is obtained from the combustion of organic materials such as wood, agricultural waste, crop residues and organic waste. It can be used to produce heat, electricity or biogas.

***Geothermal** energy: Geothermal energy is extracted from the earth's heat, which comes from the earth's core and the radioactive decay of materials. It can be used for direct heating, power generation or cooling.

***Marine** energy: Marine energy refers to energy extracted from ocean sources such as tides and waves. It is a promising renewable resource for sustainable power generation.

***Hydroelectric** power: Hydroelectric power is generated by the force of moving water, usually from dams or hydroelectric power plants.

In view of growing energy needs and the limitations of conventional sources, renewable energies offer a sustainable solution. Derived from natural resources such as solar, wind, hydro, biomass and geothermal energy, they cover a wide range of applications. Ongoing innovations cut costs, reduce emissions and promise a cleaner, safer energy future for the world. Sunlight is a clean, safe source of energy that covers the Earth's surface uniformly. This solar energy is abundant and environmentally friendly, with one hour's solar energy exceeding the world's annual consumption. Despite its potential, solar energy needs to be converted and stored to replace conventional fuels.

6- Solar energy :

Solar energy is one of the most abundant and versatile forms of renewable energy. It can be harnessed in two main ways: via photovoltaics, which directly converts sunlight into electricity using solar cells, and via solar thermal energy, which uses the sun's heat to generate electricity or for direct applications such as water heating. The enormous potential of solar energy lies in its near-universal availability and low environmental impact, although challenges remain in terms of solar panel production and efficiency.

7-History of solar energy :

Human intelligence has been harnessing solar energy for centuries, using the sun's heat to heat homes, cook food and dry crops. Simple systems such as solar ovens and solar mirrors were used in various civilizations, from the Pharaohs in Egypt to the Greeks in Greece. In the 19th century, solar energy began to be studied with the discovery of the photovoltaic effect in 1839 by French physicist Alexandre Edmond Becquerel. This discovery laid the foundations for today's solar cells, as it established the basis for the direct conversion of sunlight into electricity. Advances were made in solar technology as researchers developed more efficient solar cells for space applications. In 1958, NASA launched the first solar-powered satellite, Vanguard1. The oil crisis that began in 1973 and the global oil price crisis led the world to turn to renewable energies, including solar power. Governments invested in the research and development of solar technologies. The first residential solar systems were developed, although their adoption remained limited due to their high cost. The use of solar energy has spread to various sectors, such as rural electrification and space applications. In recent years, solar energy has grown exponentially. Production costs have fallen considerably thanks to improved technologies and increased production scale. More and more countries around the world have adopted policies encouraging the use of solar energy, making it the green energy choice.

8-Solar energy in space :

Solar resource assessment is a fundamental concept in the field of solar energy. It refers to the amount of solar energy available in a given region at a given time. It represents the solar energy incident on a land surface in a particular geographical area. This assessment is determined by factors such as latitude, altitude, surface orientation and inclination, as well as local weather conditions. Solar resource assessment is essential for determining the potential use of solar energy in a specific area. It helps to determine whether a region is suitable for the installation of solar systems, and to estimate the amount of solar energy that can be captured, whether for power generation, heating or other applications.

8.1-Le Soleil :

The Sun, the source of most renewable energy, is of vital importance to life on Earth. It provides the light and heat needed for photosynthesis, climate and ecological balance.

Sun data	Values
Star type	Spectral type G (yellow-white dwarf star)
Mass	$\approx 2 .10^{30}$ kg
Diameter	$\approx 1, 4 .10^{6}$ km (more than 100 times that of the Earth)
Surface temperature	$\approx$ 5 778 K
Core temperature	$\approx 15 .10^{60}$ C
Brightness	$\approx 3,8 . 10^{26}$ Watts
Chemical composition	(73.46%) H and (24.85%) He
Rotation time	$\approx$ 24,47 j
Estimated age	$\approx$ 4.7 billion years
Power emitted by the sun per unit area	$64 .10^{3}$ kW/m²
Total power emitted by the sun	$3.94 .10^{26}$ W $\approx$ 400 Yottawatts
Power received at Earth level per unit area	1353 $W.m^{-2}$
Distance (Sun-Earth)	150 million Km

Table 1: Sun-related data.

8.2- The sun as a source of energy :

The Sun is a thermonuclear fusion reactor that has been in operation for 5 billion years, with an estimated lifetime of $10,^{10}$ years. The process is based on the conversion of hydrogen into helium (Bethe cycle):

$$4\ {}_{1}^{1}H \longrightarrow {}_{2}^{4}He + 2\ {}_{1}^{0}e + 2\gamma + 26.7\ \ Mev$$

The sun emits large quantities of energy into space, with an estimated power of 64,000 kW/m². This radiation escapes in all directions and propagates through space in parallel beams at the speed of light, which is 3.10^{8} m/s. This collective radiation, also known as solar irradiance, travels a distance of around 150 million kilometers to reach the outer layers of the Earth's atmosphere, with a power of around 1367 W/m². This is known as the solar constant.

8.3-Atmospheric influence on solar radiation :

The influence of the atmosphere on solar radiation refers to the impact of the Earth's atmosphere on solar energy originating from the Sun as it travels through the atmosphere before reaching the Earth's surface. The atmosphere acts as a filter and modifier of solar radiation, affecting its intensity, distribution and composition figure 2.

Understanding the influence of the atmosphere on solar radiation is crucial for a variety of fields, including meteorology, climate science and renewable energy research. It also highlights the dynamic relationship between the Earth's atmosphere and the planet's energy balance.

Several factors contribute to the attenuation of solar radiation:

***Scattering:** Molecules and particles in the atmosphere scatter solar radiation, with shorter wavelengths (blue and violet) scattering more than longer wavelengths (red and orange). This is why the sky appears blue during the day.

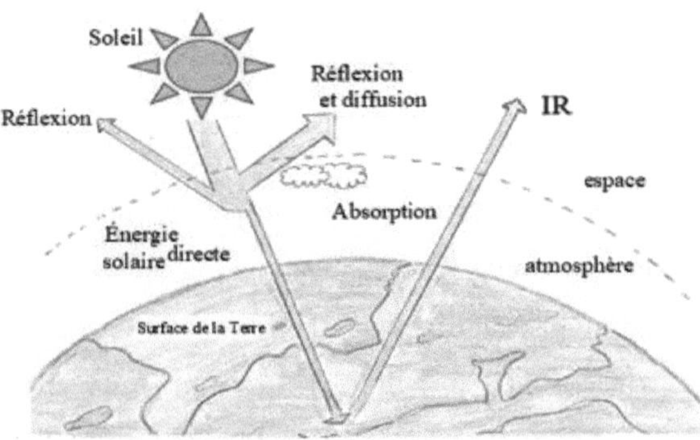

Figure 2: Atmospheric absorption.

***Absorption:** certain atmospheric components, such as gases like ozone (O_3), carbon dioxide (CO_2) and water vapor (H_2O), have the ability to selectively absorb certain wavelengths of solar radiation. For example, ozone absorbs part of ultraviolet (UV) radiation, while CO_2 and water vapor absorb specific parts of infrared (IR) radiation. This absorption can contribute to warming the atmosphere.

***Reflection:** Some solar radiation is reflected back into space by clouds, aerosols and the Earth's surface. Albedo, the reflectivity of a surface, plays a role in this reflection.

***Attenuation:** solar radiation is attenuated as it passes through the atmosphere, meaning that its intensity decreases due to scattering and absorption.

***Path length:** The angle at which solar radiation enters the atmosphere influences the length of the path it follows. Sunlight passing through a thicker part of the atmosphere is scattered and absorbed to a greater extent.

8.4-Influence of the AM air mass on solar radiation :

The concept of relative air mass (AM) in astronomy and meteorology is used to quantify the amount of atmosphere that sunlight must pass through before reaching an observer on the Earth's surface. AM varies according to the sun's angle of incidence in the sky. The equation can be used to calculate AM as a function of the sun's angle of elevation h.

$$m' = 1 / \sin(h)$$

When $h = 90^0$ (sun at zenith), $\sin(90^0) = 1$, so **m'= 1 (AM1).**

Here's a table summarizing the different cases of AM (relative air mass) as a function of the sun's angle of incidence (h):

Angle of incidence (h)	AM	Description
90^0 (**Zenit**)	AM1	Light passes through the thin atmospheric layer
30^0	AM2	Light passes through a thick layer of atmosphere
15^0	AM3	Light passes through a thick layer of atmosphere
0^0 (**Horizon**)	AM infinity	Light penetrates the great thickness of the atmosphere

Table 2: Relative air mass (AM).

Note: When the relative mass of air (m'=0), this means an observation taken from outside the Earth's atmosphere. This corresponds to the concept of AM0, representing the solar spectrum above the Earth's atmosphere.

Solar radiation attenuation refers to the reduction in intensity of solar radiation as it passes through the earth's atmosphere. The atmosphere acts as a filter for solar radiation, meaning that certain wavelengths or components of the radiation can be partially absorbed, scattered or reflected before reaching the earth's surface figure 2.

Solar radiation breaks down into approximately : Ultraviolet UV (0.20 < 1 < 0.38 mm 6.4%). Visible (0.38 < 1 < 0.78 mm 48.0%). Infrared IR (0.78 < 1 < 10 mm 45.6%).

8.5-Schematic diagram of solar radiation received at ground level :

On a clear, cloudless day, the majority of the sun's direct rays reach the earth without changing direction. When direct rays encounter clouds or impurities in the atmosphere, the result is scattered radiation that reaches the earth from all directions. Part of the incident radiation is reflected by the surroundings of the collecting surface, forming reflected radiation.

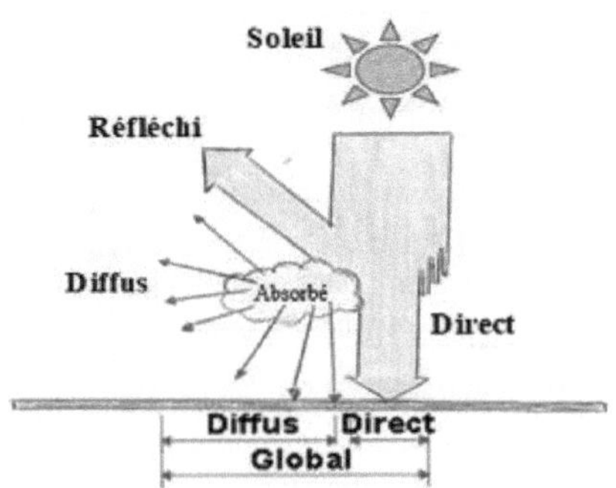

Figure 3: Global Radiation = Radiation (Direct + Diffused + Albedo).

8.5.1-Global radiation :

The sum of direct, diffuse and reflected radiation gives global radiation, which depends mainly on the season and local weather conditions. It represents the sun's total radiant energy reaching a horizontal surface on the Earth's surface in a specific unit of time. It is around 1000 W/m² for vertical solar radiation.

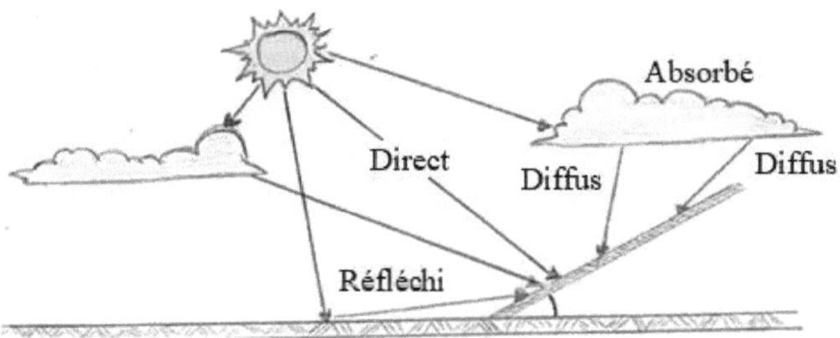

Figure 4: Direct, scattered and reflected radiation.

8.5.2-Reflected solar radiation :

The reflected component represents a smaller portion of solar radiation. The reflected part depends on the albedo, which is the solar reflection coefficient of the receiving surface. The flux intercepted by the surface then depends only on its inclination.

a-Albedo:

The earth's albedo is defined as the proportion of incident energy reflected by the earth relative to incoming energy.

Surface type	Albedo (Reflectivity)
Ocean	0.05
Moon	0.07
Forest	0.12
Concrete	0.15
Grass	0.25
Sand	0.37
Ice	0.60
Cloud	0.80
Snow	0.85
Mirror	≈1

Table 3: Albedo values.

Not all bodies have the same reflective capacity: light surfaces (such as snow, ice, etc.) have a higher albedo than dark surfaces (such as seawater, vegetation, etc.).

b-Diffuse solar radiation :

The diffuse component of solar radiation represents the flux coming from all directions across the sky. On an overcast sky, diffuse radiation would dominate direct radiation. For clear skies, the effect is less marked.

State of the sky	Description	Solar radiation value
Clear or sunny skies	Intense brightness, direct sunlight, high energy.	1,000 W/m^2 of heat
Partly cloudy skies	Partial clouds partially blocking the sun, moderate radiation.	500 W/m^2 of power
Cloudy or overcast skies	Clouds block most of the sun's direct rays and dim the light.	250 W/m^2 (in watts/m^2)
Precipitatio n	Rain or snow absorbs and scatters solar radiation, considerably reducing sunshine.	< 250 W/m^2, or

Table 4: Solar radiation in different sky conditions.

8.5.3-Direct solar radiation :

The direct component of solar radiation represents the solar flux reaching the surface directly when it is exposed to the sun. This component depends mainly on the sun's angle of elevation and the angle of incidence at a given moment.

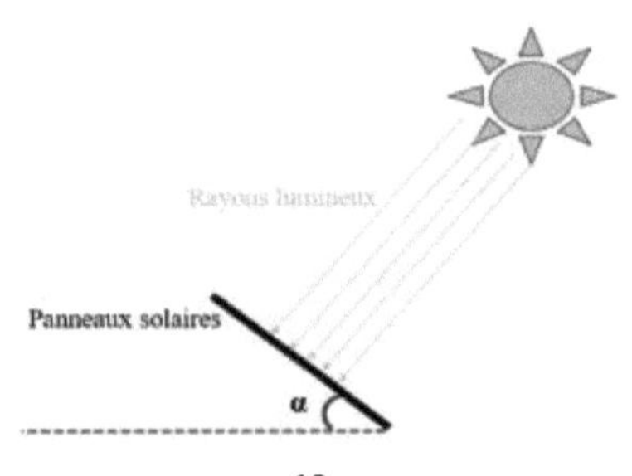

Figure 5: Direct solar radiation.

The more normal the flow is to the surface in question, the more significant it is; the more obstacles there are, the weaker it becomes.

9-Earth movements :

The Earth undergoes two main movements: rotation around its axis and revolution around the Sun.

*The Earth's rotation is the movement by which the planet rotates around its imaginary axis, called the axis of rotation. This movement is responsible for the alternation between day and night. The Earth completes one complete rotation in approximately 24 hours, which corresponds to one day. During rotation, different parts of the Earth are exposed to sunlight, creating the day-night cycle.

*The Earth's revolution is the movement by which the planet revolves around the Sun. The Earth follows an elliptical orbit around the Sun, with the Sun close to one of the foci of this ellipse. The time required for the Earth to make one complete revolution around the Sun is approximately 365.25 days, which corresponds to one year. This forms the basis of the calendar we use, with a leap year added every four years to account for the extra quarter-day.

The combination of the Earth's rotation on its axis and its revolution around the Sun results in several observable phenomena:

-Day and night: Because of the earth's rotation, different parts of the globe are exposed to sunlight, creating the day-night cycle. Because the earth is round and light travels in a straight line, the sun cannot illuminate the entire surface of the earth at the same time. When one side of the earth is illuminated, it's day, and if the other side receives no sunlight, it's night.

-Seasons: Due to the tilt of the Earth's axis in relation to its orbital plane, different regions of the Earth receive varying amounts of solar radiation throughout the year. This leads to seasonal changes, such as summer, winter, spring and autumn.

-Leap years: To compensate for the extra quarter day in the year, a leap year is added every four years.

9.1-Rotation and orbit of the Earth around the Sun :

In its orbit around the Sun, the Earth follows an elliptical trajectory, completing a full orbit around the Sun in one year and a full rotation on its axis in 24 hours. The axial inclination, with a declination angle of 23°27', is responsible for the changing of the seasons.

***During the summer solstice (June 21):** the Earth is tilted towards the sun's rays. A person living at a latitude of 66°33' N would have to stay awake until midnight to see the Sun rotate around the north, dip to touch the horizon and start rising again towards the eastern part of the sky. The height of the Sun at solar noon is 23°27' higher than at equinox.

$$H = 90° - L + 23°27$$

***During the winter solstice (December 22):** the angle of inclination is reversed, and it's the Tropic of Capricorn (latitude 23°27' S) that receives perpendicular sunlight. The Sun's height at solar noon is 23°27' lower than at equinox.

$$H = 90° - L - 23°27'$$

***During the spring and autumn equinoxes (March 21, September 21):** at noon, the Sun's rays are perpendicular to the equator, and everywhere on the globe, days and nights are of equal length. The height of the Sun at solar noon is the easiest to calculate.

$$H = 90° - L$$

10-Location of a site on the Earth's surface :

It is possible to determine the position of the sun in the celestial sphere as a function of time and the position of the observer on Earth. To locate a precise site on the Earth's surface, the following parameters are defined:

10.1-Geographical coordinates :

a-Latitude (θ): Specifies the location of a point relative to the equator; it ranges from 0° at the equator to 90° North (or South) at the poles. It represents the angular distance of site S from the plane of the equator, with θ ranging from -90° to +90°.

Thus: $\theta > 0$ indicates north, $\theta < 0$ indicates south.

b-Longitude (φ): This is the angle φ formed between the Greenwich meridian and the site meridian. Longitude varies from -180 (west) to +180 (east). Since the Earth takes 24 hours to complete one rotation (360°), each hour represents a longitude difference of 15°, making each degree of longitude equal to 4 minutes.

c-Altitude (z) : This is the vertical distance, expressed in meters, between the point of relief on the earth's surface and sea level, taken as the reference surface.

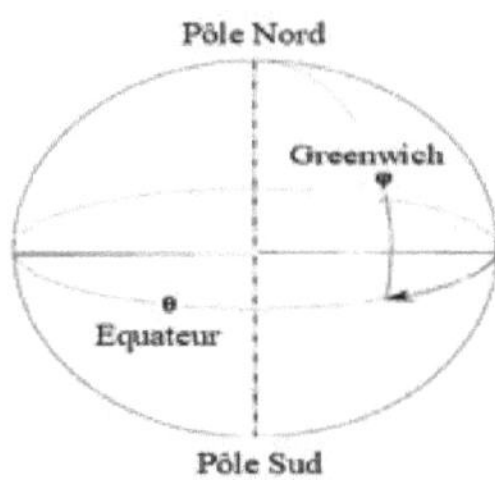

Figure 6: Geographical coordinates.

11-Apparent motion of the Sun :

To fully understand and use the Sun's influence in the choice and treatment of a site, it's essential to know the Sun's position in the sky at all times. This information is crucial for calculating solar gain, choosing the orientation of a building, positioning active solar systems (thermal or photovoltaic), designing the surrounding outdoor spaces, providing natural lighting for interior rooms, determining window locations, landscaping, etc.

Figure 7: Apparent motion of the Sun.

The Sun's position in the celestial sphere is determined at any time of day using two coordinate systems:

11.1-Equatorial coordinates :

Equatorial coordinates are independent of the observer's location on Earth, but are linked to the moment of observation. The Sun's position is expressed using two angles, which are :

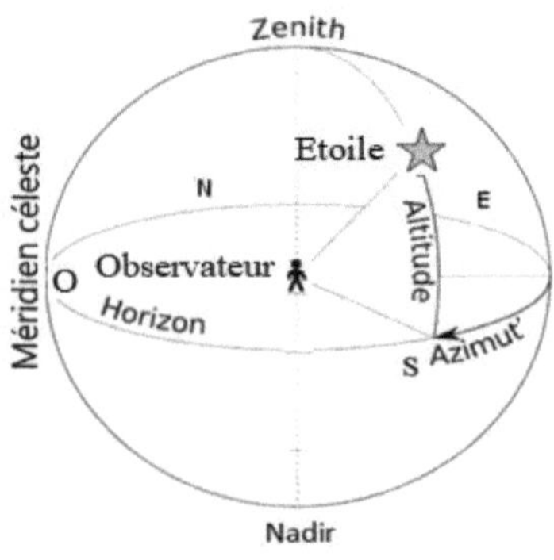

Figure 8: Ground coordinates.

a-Declination (δ) :

It's the angle between the direction of the Sun and the equatorial plane, and varies sinusoidally throughout the year. Several expressions have been developed to calculate declination, for example.

$$\delta = 23.45 \times \sin\left[360/365 \times (n+284)\right]$$

Where n is the number of the day of the year from 1 to 365.

Spring equinox: March 21; $\delta = 0$

Summer solstice: June 22; $\delta = +23° 27'$.

Autumn equinox: September 23; $\delta = 0$

Winter Solstice: December 22; $\delta = -23° 27'$.

As declination is a sinusoidal function, it evolves rapidly around the equinoxes (0.4°/day), while remaining virtually stationary during the periods surrounding the summer and winter solstices.

b- Hour angle (ω) :

The hour angle measures the Sun's movement relative to noon, the moment when the Sun crosses the meridian plane of the zenith. This angle is formed between the projection of the Sun on the equatorial plane at a given instant and the projection of the Sun on the same plane at true noon. The hour angle is given by the following equation:

$$\omega = 15 \times (TSV - 12)$$

Where ω in degrees and TSV is true solar time in hours as described below.

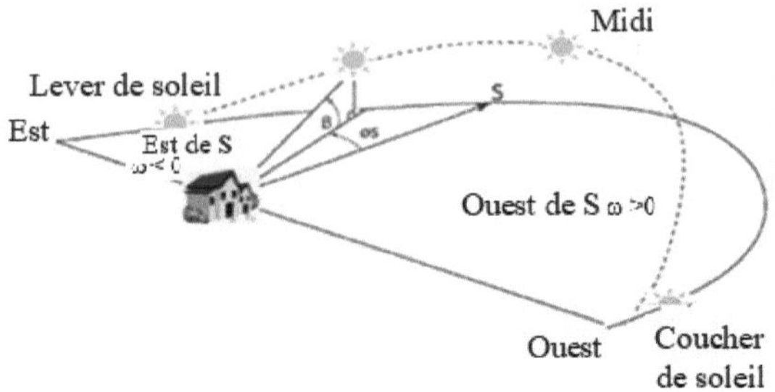

Figure 9: Solar hour angle (ω).

11.2-Horizontal coordinates :

The Sun is identified by the following quantities:

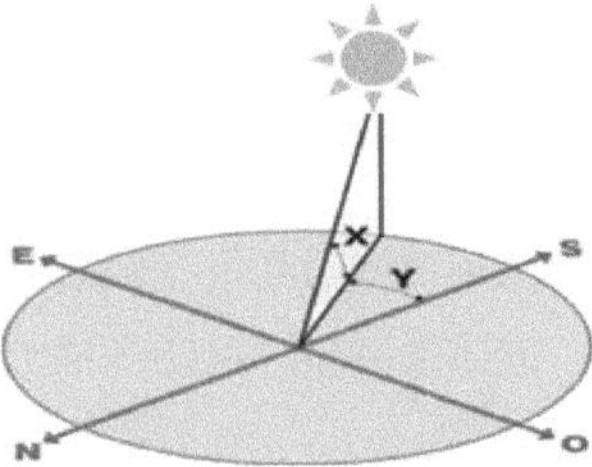

Figure 10: Solar coordinates in the horizontal coordinate system.

a-Solar elevation angle (h): This is the angle between the direction of the Sun and the horizontal plane. h varies from 0° to 90° towards the zenith and from 0° to -90° towards the nadir. The sun's angle of elevation is given by :

$$\sin h = \sin \theta \times \sin \delta + \cos \theta \times \cos \delta \times \cos \omega$$

b-Azimuth angle (a): The azimuth angle is the angle between the vertical plane containing the solar beam and the south direction. It is measured from 0° to 360° clockwise from the south. The equation for calculating the azimuth angle is given below:

$$\sin a = \cos \delta \times \sin \omega / \cos h$$

c-Zenith angle (Z): This angle lies between the direction of the Sun and the vertical at the location (zenith). The angle Z is complementary to h.

12-Time :

The Earth undergoes two types of movement: its rotation around its polar axis and its revolution around the Sun. The Earth's rotation on itself defines the notion of the solar day. A complete rotation takes 24 hours, thus defining time, since each hour corresponds to an angular deviation of 15°. The Earth's revolution around the Sun defines the seasons and gives rise to the distinction of true solar time (TSV).

$$\omega = 15\times (TSV - 12)$$

Where ω in degrees and TSV is true solar time (hour).

12.1- True solar time (TSV) :

Solar time (TSV) is defined by the Sun's true angular coordinates. It is calculated from mean solar time (MST) by adding the equation of time (Et) :

$$TSV = TSM + Et$$

12.2-Average solar time (TSM) :

The difference between mean solar time and universal time is called longitude correction. Mean solar time (MST) is related to universal time (UT) by the following relationship :

$$TSM = TU + \varphi/15$$

Where φ represents the longitude of the location, positive for east longitudes and negative for west longitudes. SST is expressed in hours.

12.3-Universal Time (UT) :

Universal Time (UT) is determined by the time at which the Sun crosses the prime meridian. The chosen prime meridian is Greenwich Mean Time (GMT), and Mean Solar Time (MST) corresponds to Universal Time (UT) at longitude 0°.

12.4-Time equation (Et) :

The equation of time takes into account the variation in the Earth's rotation speed. It is given by the following equation :

$$Et = 9.87\times \sin(2\times\beta_0) - 7.53\times\cos(\beta_0) - 1.5\times\sin(\beta_0) \text{ (minutes)}$$

Where:

β_0 : the angle according to the number of the day in the year. $\beta_0 = 360/365\times(n-81)$ (degrees)

Where Et is expressed in minutes and n is the number of the day of the year starting on January 1st.

Note: For a location of longitude φ, there is a direct correspondence between hour angle ω, solar time TSV, local solar time TSL and universal time TU :

$$\omega = 15\times (TSV - 12)$$

$$TSV = TSM + Et$$

$$TSM = TU + \varphi /15$$

$$\omega = 15\times (TU + \varphi /15 + Et - 12)$$

12.5-Legal time TL :

Legal time, abbreviated TL, refers to the time displayed by clocks and watches in a specific region or country. It serves as a reference time used by the population to coordinate daily activities, work schedules, appointments and other events. Legal time (TL) is generally determined on the basis of Coordinated Universal Time (UTC), also known as civil time or standard time. Coordinated Universal Time is a globally recognized time scale based on atomic clocks and used to maintain a common time reference around the world. The relationship between TL and TU depends mainly on time zones and any adjustments for summer and winter time, as well as seasonal correction.

$$\mathbf{TL = TU + C + C_{12}}$$

$$\mathbf{TU = TL - C_1 - C_2}$$

C_1 : Time zone correction ($C_1 > 0$ East of GMT, West of GMT $C_1 < 0$).

C_2 : Season correction ex (C_2 =0h in Winter and C_2 =+1h in Summer).

For example: France C_1 =+1h and C_2 =0 for winter and C_2 =+1h for summer.

The various transitions from one hour to the next are summarized below:

TL-C_1-C_2- φ/15+Et

TU

TSM

TSV

The equation of time varies from -14.5 minutes (February 10 to 15) to +16.5 minutes (October 25 to 30).

12.6-Day number in year n : Calculating the number of the day in the year involves adding the number of the day of the month (day of the month) to the characteristic number of each month. n varies from 1 (January 1) to 365 (December 31), i.e. 366 for a leap year. The table below shows the characteristic numbers for each month.

Month	Jan	Feb	Mar	Apr	May	Jun	Jul	August	Sep	Oct	Nov	Dec
Number of the day	0	31	59	90	120	151	181	212	243	273	304	334

Table 5: Day number at the beginning of each month.

12.7-Sunrise and sunset :

Given the declination δ and latitude θ of the observed location, we can calculate the actual solar time of sunrise and sunset using the following equations:

$$\mathbf{TSV_{Sunrise} = 12 - [Arc\cos(-\tan(\theta) \times \tan(\delta))] / 15.}$$

$$\mathbf{TSV_{sunset} = 12 + [Arc\cos(-\tan(\theta) \times \tan(\delta))] / 15.}$$

Note: If $\tan(\theta) \times \tan(\delta)>1$, the sun does not rise;

If $\tan(\theta) \times \tan(\delta)<1$, the sun does not set.

12.8-Sunshine duration :

The sunshine duration of a day, noted Di, corresponds to the period between sunrise and sunset in the absence of clouds. This duration is equivalent to the astronomical length of the day, and can easily be obtained in terms of hourly angle using the following equation:

$$\mathbf{D_i = (2/15) \times [Arc\cos(-\tan(\theta) \times \tan(\delta))]}$$

13-Shadow phenomenon:

Figure 11: Shadow phenomenon

Any obstacle that can block the sun's rays is called a solar mask. Generally speaking, there are two types of masks:

***Close masks**: correspond to nearby obstacles (vegetation, buildings, etc.).

***Far shadows**: correspond to distant obstacles, such as mountains and hills on the horizon. The shading effect at a given moment depends on the sun's coordinates and the geometric characteristics of the shading system or obstacle, in relation to the shaded or protected element. To make the best possible use of the sun's rays, obstacles to solar radiation or solar masks should be avoided wherever possible.

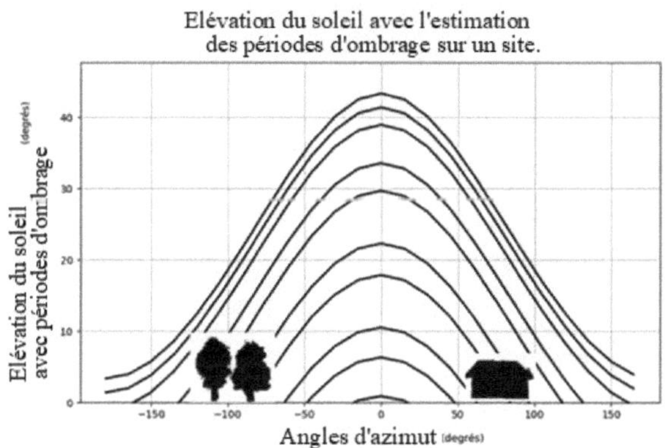

Figure 12: Diagram of the sun's path during periods of shading on a site.

Not only do sun path diagrams, like the one shown in Figure 11, help us understand where the sun is at any given time, they also have a very practical application in the field when trying to predict shading patterns at a site.

14-Energy measurement of radiation :

The difference between "Solar Irradiance" and "Solar Irradiation":

***Solar irradiance:** Solar irradiance measures the power of solar radiation striking a specific surface per unit area, expressed in watts per square meter (**W/m²**). It represents an instantaneous measurement of the power of solar radiation at a given moment.

***Solar irradiance:** Solar irradiance measures the total amount of solar energy received by a specific surface over a given period (day, hour, month, year), expressed in joules per square metre (**J/m²**) or watt-hours per square metre (**Wh/m²**). It represents a cumulative measure of solar energy over a specified period of time.

15-Components of solar irradiation :

Solar radiation at ground level consists mainly of direct radiation, coming directly from the sun and slightly weakened by scattering or absorption as it passes through the atmosphere, and diffuse radiation, coming from the whole sky due to the scattering of direct radiation by molecules and aerosols... Thus, the overall radiation received at ground level is the sum of these two components:

Solar irradiance Instantaneous flux **(W.m)$^{-2}$**	GLOBAL G* DIFFUSE D* DIRECT I*	**G* = I*+D***
Solar irradiation Energy received over a period of time **(KWh.m)$^{-2}$**	GLOBAL G DIFFUSE D DIRECT I	**G = I+D**

Table 6: Notations for solar radiation components.

16-Solar radiation meters :

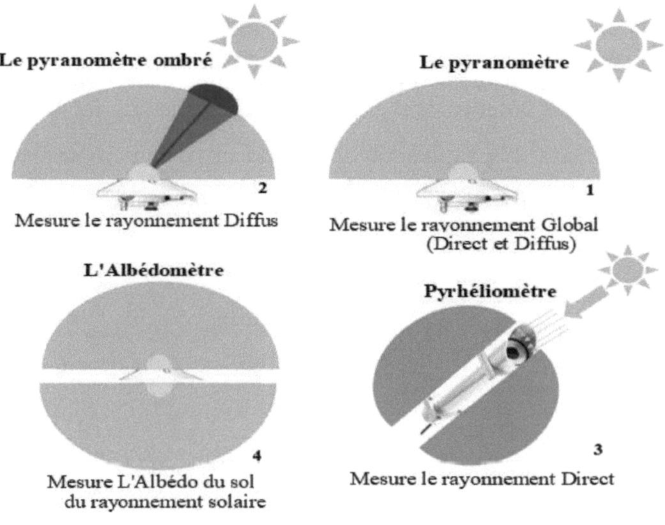

Figure 13: Solar radiation measuring instruments.

17-World Solar Potential :

The world's solar potential is immense and unevenly distributed, offering an abundant and almost unlimited source of renewable energy. Around 173,000 terawatts of solar energy reach the Earth every day, far exceeding the world's current energy needs. However, this potential varies considerably according to latitude, climate, altitude and geographical orientation. Regions close to the equator, such as sub-Saharan Africa, the Middle East and parts of Latin America, benefit from intense and constant solar radiation, making them particularly favorable zones for harnessing solar energy. Conversely, regions further from the equator, such as Northern Europe and Canada, have less solar potential due to low incidence of solar radiation and often cloudy weather conditions. Despite these disparities, technological advances and the falling cost of photovoltaic systems mean that solar energy can now be harnessed efficiently even in regions with lower solar potential, making this energy source increasingly accessible worldwide. This form of energy offers many advantages in terms of thermal conversion, mainly for heating and power generation. It's an available, cost-effective and environmentally-friendly energy source that requires minimal maintenance.

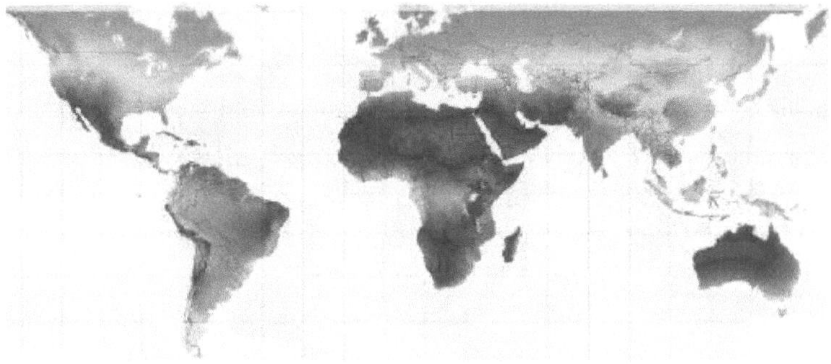

Figure 14: World map showing average annual sunshine.

Photovoltaic solar energy

1-Introduction :

As industrialization progresses, the growing problem of energy scarcity is gaining in importance. Photovoltaic (PV) technologies are developing rapidly, and occupying an increasingly important place in the field of electrical technology. They are positioning themselves as the main source of green energy for the new century. Major efforts are currently underway to improve the electrical performance of PV cells (or modules) and systems, with the aim of reducing energy losses in photovoltaic installations. This initiative aims to significantly reduce the costs associated with PV installations, while encouraging citizens to adopt more green energy.

2-History of photovoltaic energy :

Knowledge of the photovoltaic effect goes back a very long way, and the stages in its historical development are as follows:

In 1839, French physicist Edmond Becquerel discovered the process of using sunlight to generate electric current in a solid material, known as the photovoltaic effect.

In 1875, Werner Von Siemens presented a paper on the photovoltaic effect in semiconductors to the Berlin Academy of Sciences. However, until the Second World War, this phenomenon remained mainly a laboratory curiosity.

In the 1950s, researchers at the Bell Telephone Company in the USA succeeded in manufacturing the first solar cell, the primary component of a photovoltaic system.

In 1954, three American Bell Labs researchers, Gerald Pearson, Daryl Chapin and Calvin Fuller, developed a high-efficiency photovoltaic cell to meet the growing needs of the space industry for innovative solutions to power its satellites.

In 1958, a cell with an efficiency of 9% was developed. The first satellites powered by solar cells are sent into space.

In 1973, the first house powered by photovoltaic cells was built at the University of Delaware.

In 1983, the first car powered by photovoltaic energy covered a distance of 4,000 km in Australia, marking a significant advance in the practical application of this technology.

3-Photovoltaic conversion :

3.1-Definition :

The conversion of photon energy from sunlight into electricity is based on the fundamental principle of the photovoltaic effect.

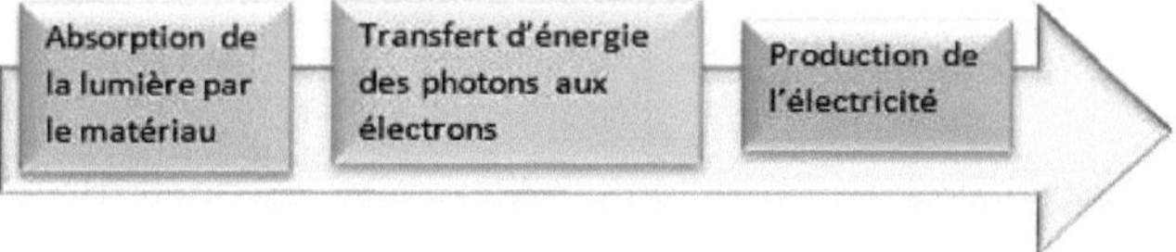

3.2-A little history of the photovoltaic effect (origin of the discovery) :

In 1839, the discovery was made of a phenomenon that generates a small amount of electricity when certain materials are exposed to light, known as the photovoltaic effect. The term "photovoltaic" comes from "photo" (from the Greek "phos" meaning "light") and "Volt" (in homage to the physicist Alessandro Volta, a major contributor to electrical research).

Figure 15: Alexandre Edmond Becquerel (1820 - 1891)

This breakthrough has its roots in the work of Alexandre Edmond Becquerel, a French physicist who was born in Paris on March 24, 1820 and died on May 11, 1891. He is famous for his discovery of the photovoltaic effect in 1839 and for producing the first color photograph in 1848. The photovoltaic effect, discovered by Edmond Becquerel, involves the transmission of light energy to the electrons of a semiconductor, forming a photovoltaic cell without mechanical intervention, noise, pollution or the need for fuel.

The Prix Becquerel is named after Alexandre Edmond Becquerel, a pioneer of photovoltaic technology. Born in Paris on March 24, 1820, he succeeded his father Antoine César Becquerel at the Muséum National d'Histoire Naturelle. At the age of 19, Becquerel developed an electrochemical actinometer consisting of two photosensitive plates coated with silver chloride, placed in separate compartments filled with acidified water. While one of the plates was exposed to light, the other remained in the dark. Using a galvanometer, a small electric current resulting from exposure of the first plate to light was measured, demonstrating the photovoltaic effect. Becquerel concluded that light, not temperature, was the cause of this phenomenon.

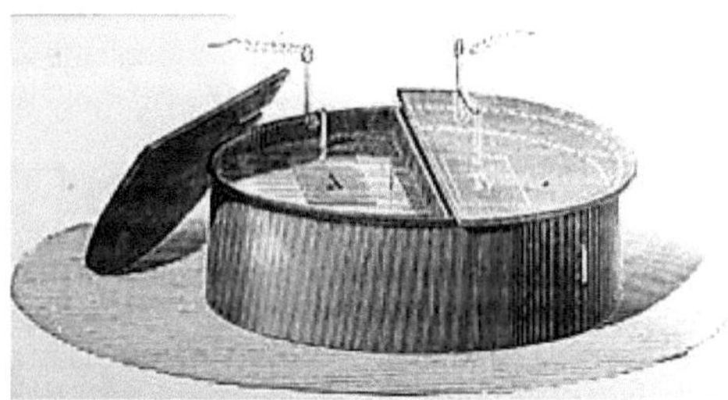

Figure 16: Electrochemical actinometer

In addition to his work on photovoltaics, Alexandre Edmond Becquerel was actively involved in other research related to the effects of light, including photography and phosphorescence. In 1868, he published the important book "Light, its causes and effects".

It's important to note that Alexandre Edmond Becquerel should not be confused with his son Henri Becquerel, winner of the Nobel Prize in Physics for the discovery of radioactivity.

4-Photovoltaic effect :

Edmond Becquerel was a remarkable innovator in the study of the effects of light in various scientific fields. Albert Einstein explained the photoelectric phenomenon in 1905, but it wasn't until the early 1950s that this led to the creation of the first silicon photovoltaic cells. The first silicon homojunction solar cells, developed in the 1950s, marked the beginning of photovoltaic conversion, with an initial efficiency of around 4.5%. From the 1960s onwards, advances enabled efficiencies in excess of 10% to be achieved with monocrystalline silicon-based cells, although their use was initially limited to space due to their high cost. Since then, the photovoltaic industry has focused its efforts on improving performance and reducing costs, enabling solar energy to be more widely and competitively adopted on the market.

5-The solar cell :

5.1-Definition :

The photovoltaic solar cell converts light directly into electricity using semiconductor materials such as silicon. It consists of a semiconductor layer, an anti-reflective layer, a conductive grid (cathode) on top, a conductive metal (anode) underneath, and sometimes reflective multilayers to improve efficiency. Silicon, used for its availability and non-toxicity, comes in two main forms: monocrystalline and multicrystalline. Despite its disadvantage of having an indirect gap of 1.1 eV, silicon is widely used and continually improved to increase solar cell efficiency. Researchers are also exploring other materials with different energy gaps to further improve these technologies.

Symbole chimique	Propriété	Valeur
Si	**Numéro atomique**	14
	Masse atomique	28.085 u
	Groupe	14 (Famille du Carbone)
	Période	3
	Structure électronique	$1s^2\ 2s^2\ 2p^6\ 3s^2\ 3p^2$
	Configuration électronique	$[Ne]\ 3s^2\ 3p^2$
	Masse volumique	$2330\ Kg/m^3$
	État de la matière	Solide à 25^0C
	Point de fusion	1 683 °C
	Point d'ébullition	2 628 °C
	Coefficient de dilatation thermique	$4,2\ .10^{-6}/K$
	Type de matériau	Semi-conducteur
	Electronégativité	1.90
	Abondance (la croûte terrestre)	27.7%

Table 7: Silicon (Si) characteristics

Silicon is a chemical element of the semiconductor metal family, with the symbol Si and atomic number 14. It belongs to group 14 of the periodic table. Silicon is the second most abundant element in the earth's crust, after oxygen, and is widely used in various fields, particularly in the electronics industry.

The chemical formula SiO_2 stands for silicon dioxide, also known as silica. It is a chemical compound composed of one silicon atom bonded to two oxygen atoms. Silica is one of the most common forms of silicon in nature, and occurs in a variety of crystalline and amorphous forms.

Semiconductor	Symbol	Eg (eV)(gap)
Silicon	If	1.1 (indirect)
Germanium	Ge	0.7(indirect)
Gallium arsenide	GaAs	1.4 (direct)
Gallium nitride	GaN	3.4 (direct)
Cadmium sulfide	CdS	2.42(direct)

Table 8: Gap values for various semiconductors

6-Semiconductor energy band theory :

6.1-Electronic structure of atoms: Energy band theory begins with an understanding of the electronic structure of atoms. Each atom has discrete energy levels, but when many atoms combine to form a solid material, these energy levels merge to form continuous energy bands.

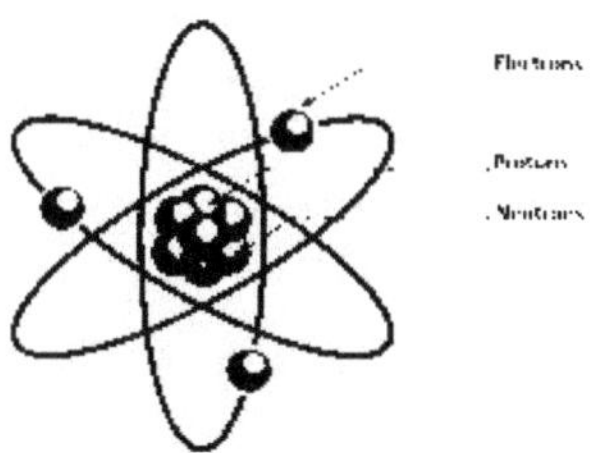

Figure 17: Electronic structure of the atom

6.2-Formation of energy bands : When a large number of atoms are grouped together in a crystal, the energy levels of individual electrons overlap, creating energy bands. The lowest energy band is the "valence band", occupied by electrons bound to atoms, while the highest energy band is the "conduction band", where electrons can move freely.

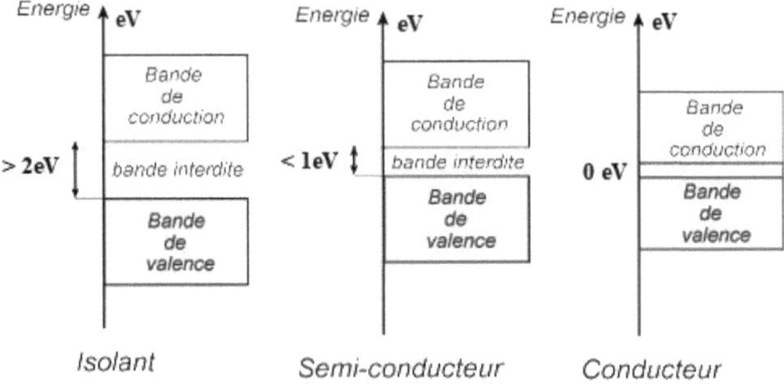

Figure 18: Energy bands

a-Band gap : Between the valence band and the conduction band lies a forbidden region (gap or band gap) where energy levels are not allowed. This band gap determines the material's electrical conductivity.

- ***Direct gap:*** When the minimum of the conduction band and the maximum of the valence band correspond to the same value of the wave vector (k), the gap is direct. GaAs and CdTe are examples of direct-gap materials.

- ***Indirect gap:*** This time, the transition between the band extrema is not vertical, but oblique. At energies equal to or slightly greater than the gap energy, the photon can only be absorbed by a phonon. This adds a new condition to absorption, greatly reducing its probability. Crystalline silicon is an example of an indirect-gap semiconductor.

Note: When the photon's energy is lower than the material's gap, the transition is not possible and the photon is not absorbed.

b-Isolators : Insulators have a larger band gap, requiring a significant amount of energy to move an electron into the conduction band. At room temperature, they have very low electrical conductivity.

c-Semiconductors: Semiconductors have a small band gap, which means it takes relatively little energy to move an electron from the valence band to the conduction band. At room temperature, some electrons can be thermally excited into the conduction band, enabling electrical conduction.

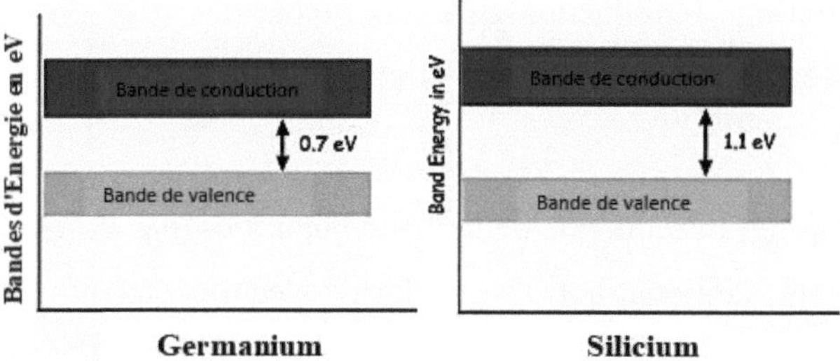

Figure 19: The Ge and Si semiconductor gap

d-Conductors: Conductors have a conduction band that partially overlaps with the valence band, allowing electrons to move around easily.

7-Doping :

Doping is the technique used to transform an intrinsic semiconductor, which is a pure material, into an extrinsic semiconductor, by adding specific impurities. This process modifies the electrical properties of the material by increasing the number of charge carriers, either electrons or holes, thus improving its conductivity and enabling its use in various electronic devices.

7.1-Intrinsic semiconductors :

An intrinsic semiconductor is a pure semiconductor material, with no intentional impurities added. At room temperature, electrons can thermally gain enough energy to move from the valence band to the conduction band, creating electron-hole pairs. The number of free electrons is equal to the number of holes.

Examples: pure silicon (Si), pure germanium (Ge).

7.2-Etrtrinsic semiconductors :

An extrinsic semiconductor is a semiconductor material that has been doped with impurities to modify its electrical properties.

N-type doping: Introduction of donor impurities (e.g. phosphorus in silicon) which provide additional electrons, increasing the free electron density.

P-type doping: Introduction of acceptor impurities (e.g. boron in silicon) which create additional holes by capturing electrons, increasing the hole density.

Example: Silicon doped with phosphorus (N-type) or boron (P-type).

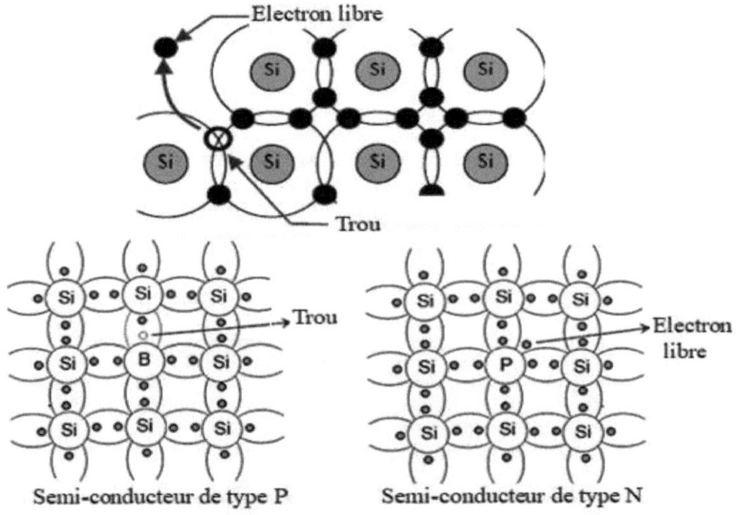

Figure 20: An extrinsic semiconductor

7.3 P-N junction :

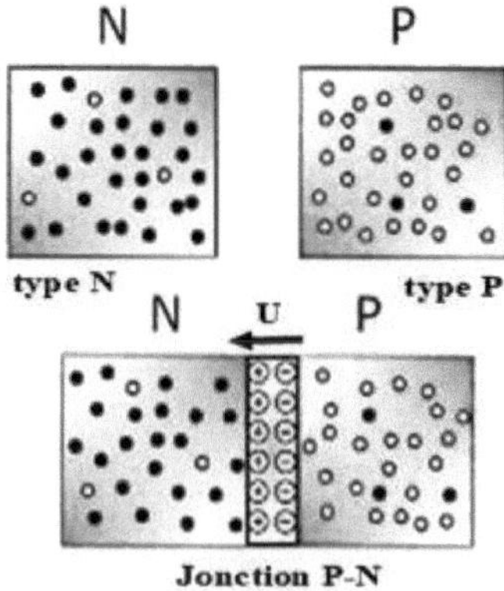

Figure 21: The PN junction in the semiconductor

8-Solar cell operation :

The solar cell converts sunlight into electricity using the photovoltaic effect. Light photons striking a cell with a doped P-N junction excite the electrons in the semiconductor material, creating electron-hole pairs. This process induces a movement of charges across the junction, generating a direct current. Metal wires collect this current and direct it to an external circuit, powering appliances.

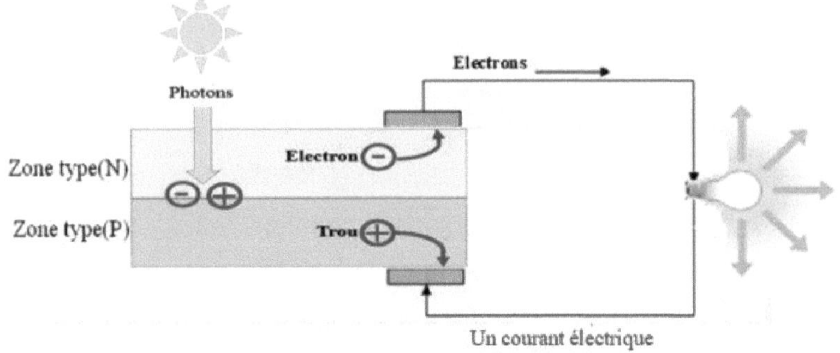

Figure 22: How a solar cell works

9-Solar cell construction :

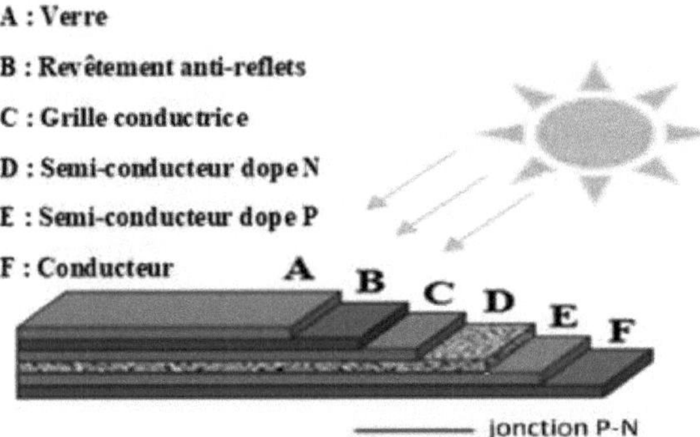

Figure 23: Construction of a solar cell

A: Glass: Used as a protective and support layer for the solar cell.

B: Anti-reflective coating: Reduces light reflection at the cell surface, thus increasing light absorption.

C: Conductive grid: Distributes the electric current generated by the solar cell.

D: N-doped semiconductor: Semiconductor material with impurities to create a negative charge.

E: P-doped semiconductor: Semiconductor material with impurities to create a positive charge.

F: Conductor: collects the electric current generated by the solar cell and routes it to an external electrical circuit.

10-Photovoltaic system components :

10.1-Cell, panel and photovoltaic field :

Cell groups :

-The photovoltaic cell is the basic unit for converting light energy into electrical energy.

-A photovoltaic panel is an assembly of photovoltaic cells. Sometimes panels are also called photovoltaic modules.

-When several panels are grouped together on the same site, the result is a photovoltaic field.

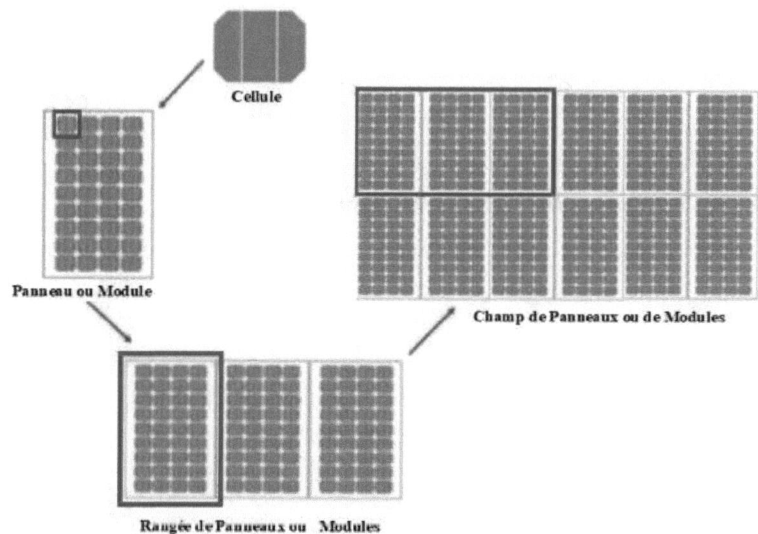

Figure 24: Cell, panel and photovoltaic field

10.2-Composition of a photovoltaic panel :

Traditional solar panels are made up of several laminated layers, forming a complex structure designed to effectively protect the photovoltaic cells from external aggression. Here's an overview of the key components of this structure:

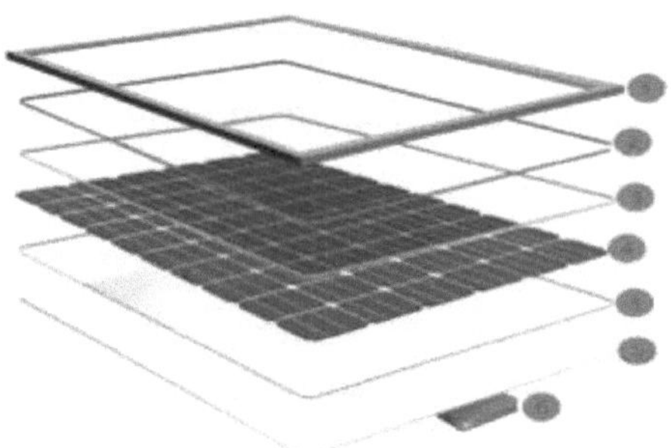

Figure 25: Composition of a photovoltaic panel

***Aluminium frame :**

Reinforces the entire solar panel. Facilitates handling, fixing and ensures panel stability.

***Tempered Glass:**

Protects solar cells against impact (e.g. hail). Maintains the transparency needed to capture sunlight.

***EVA (Ethylene Vinyl Acetate) Encapsulation Film:**

Hermetically encapsulates photovoltaic cells. Prevents air and moisture from reaching the cells, providing long-lasting protection against solar radiation.

***Photovoltaic Solar Cells :**

Convert solar radiation into electricity.

***Tedlar membrane:**

Protects against external aggressions such as corrosion and temperature variations. Ensures greater panel durability in a variety of environmental conditions.

*The **Junction Box :**

Connects photovoltaic cells to the rest of the electrical installation. Essential for integrating the panel into the overall electrical system.

11-The different types of solar cells :

Photovoltaic technologies have made significant advances over the last thirty years, marked by laboratory records, including a conversion efficiency of 41.6% achieved in the USA in 2008. This underlines the potential for further improvement, despite the theoretical efficiency limit, known as the Shockley-Queisser limit, estimated at around 33.7% for a single solar cell. Continued progress suggests the possibility of exceeding this limit, although technical challenges remain for large-scale applications. Worldwide production of solar cells is mainly divided into three technologies: crystalline silicon, thin-film and organic cells. These technologies have coexisted on the market for several years, each offering varying prices and efficiencies, and generating growing interest in performance enhancement.

There are several types of solar cell, each with its own efficiency and cost. A table summarizing the main types of solar cells:

Cell type	Features	Benefits	Disadvantages	Uses
Silicon Monocrystalline	- Very good yield: 15 to 20%. -Output: approx. 150 Wp/m². - Long service life (30 years).	- Raw material widely available.	- Low yield in low light conditions. - Loss of efficiency as temperature rises. - High manufacturing costs.	- Low-power devices, space applications.

Polycrystalline silicon	- Good performance : 14 à 18 %. -Power: approx. 100 Wp/m². - Long service life (30 years).	- Cheaper to manufacture than monocrystallin e.	- Low yield in low light and low yield in full sun. - Loss of efficiency as temperature rises.	- Generators of all sizes, both grid-connected and off-grid.
Amorphous silicon	- Low yield: 5-9%. -Output: approx. 60 Wp/m². - Fairly long service life (10 years).	- Low manufacturing costs.	- Low overall yield. - Low-light operation. - Not very sensitive to high temperatures.	- Low-power devices, energy production (solar calculators and watches).
Silicon-free thin-film cell CIS(copper, indium, selenium) / CIGS(copper, indium, gallium and selenium).	- Yield : 9 à 11%	Better yields than other thin-film photovoltaic cells -The cell can be built on a flexible substrate.	Thin-film cells are less efficient than thick-film cells	

Table 9: Main solar cell types

Other technologies :

Perovskite solar cells boast laboratory efficiencies in excess of 19%, thanks to their ability to absorb a wide range of wavelengths, making them efficient even in low light conditions. Their manufacture uses readily available materials and low-cost techniques, promising to reduce solar energy costs.

The future of perovskite solar cells is bright. Their flexibility and high efficiency mean they can be integrated into everything from building facades to portable devices. Perovskite could play a crucial role in the transition to a sustainable energy future, transforming the solar energy landscape thanks to its outstanding properties and ongoing potential for improvement.

12-Photovoltaic module assembly :

For solar panels to function as part of a photovoltaic system, several modules are needed, connected to each other. There are two ways of connecting them: in series or in parallel. These two options are quite different, and one or the other is used according to need.

12.1-Series combination of solar cells :

Series grouping is a configuration that increases the output voltage in a system composed of n cells connected in series. The output voltage, denoted Us, can be expressed in general terms as follows:

$$Us = n \cdot Uc$$

Here, Uc represents the voltage supplied by an individual cell. It's important to note that, in this type of grouping, the current is common to all cells. This means that the same current flows through all cells connected in series. This principle is essential in many applications where a higher voltage is required.

Example: grouping of 3 cells in series.

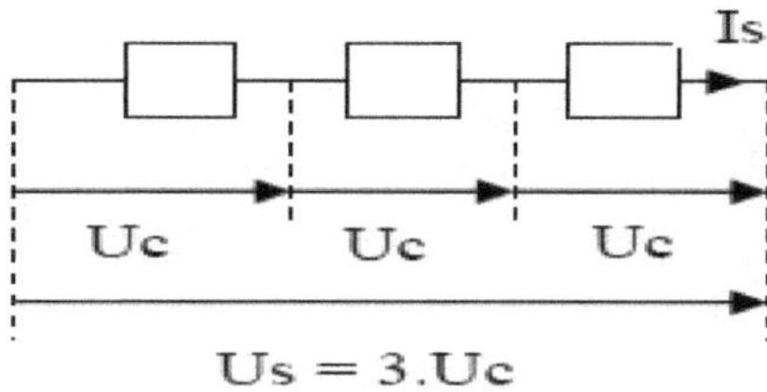

Figure 26: Grouping of cells in series

Is is constant , Us=3.Uc

Ex: Uc=3volt , Us=3.3=9 volts.

12.2-Parallel association of solar cells :

Parallel grouping is a configuration designed to increase the output current in a system comprising n cells connected in parallel. The output current, denoted Is, can be expressed in general terms by the following relationship:

$$Is = n \cdot Ic$$

Here, Ic represents the current supplied by an individual cell. It's important to note that, in this type of grouping, the voltage is common to all cells. This means that all cells connected in parallel are subjected to the same voltage.

Example: grouping of 3 cells in parallel.

In this case, the output current Is would be three times the current Ic supplied by a single cell.

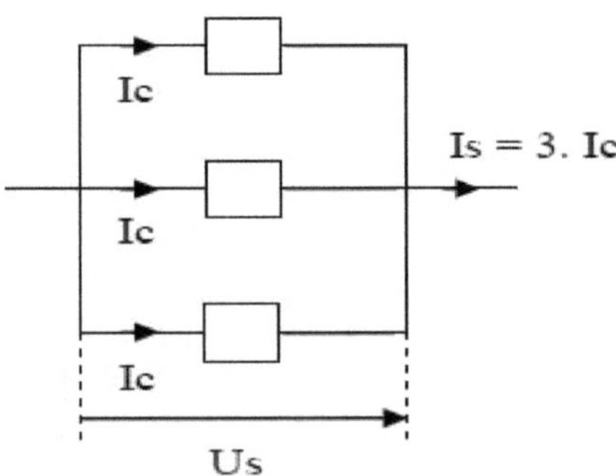

Figure 27: Parallel cell grouping

Us is constant , Is=3.Ic

Ex : Ic=0.5A , Is=3.0.5=1.5 A .

13-Electrical characteristics of a cell :

The electrical characteristics of a cell, particularly a photovoltaic (solar) cell, describe the relationship between voltage, current and power generated by the cell under different conditions. These characteristics are essential for evaluating the performance of a solar cell in real-life applications. The main electrical characteristics include the current-voltage characteristic (I-U), the power-voltage characteristic (P-U), and the maximum power (Pmax).

13.1-Cell characteristics (current-voltage) :

The current-voltage characteristic shows the relationship between the current produced by the solar cell and the voltage across its terminals under different illumination and temperature conditions. Typically, the I-V characteristic is represented as a curve, showing how current varies as a function of voltage.

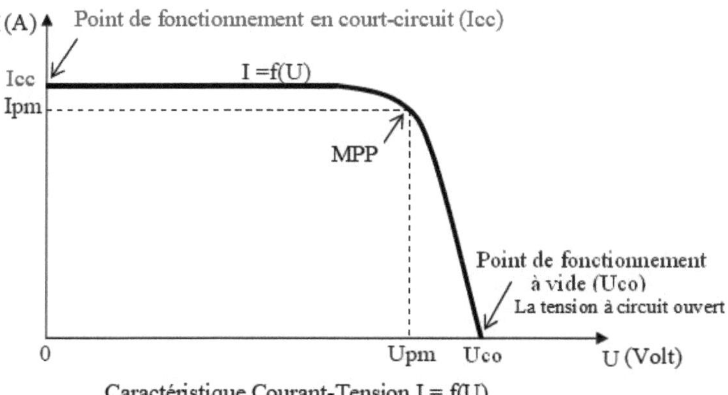

Figure 28: Current-voltage characteristic of the solar cell

There are 3 main points in this curve:

-Short-circuit current (Icc) :

The short-circuit current (Icc) is the current flowing through the solar cell when the terminal voltage is zero (short circuit). This is the maximum current the cell can deliver.

-Open circuit voltage (Uoc) :

The open-circuit voltage (Uoc) is the voltage across the solar cell when the current is zero. It's the maximum voltage the cell can produce.

-Maximum power (MPP) :

Maximum power, Pmax, is the maximum power the solar cell can deliver. It is obtained at the operating point where the product of current Ipm and voltage Upm is highest, i.e. on the P-U characteristic at the MPP (Maximum Power Point) point.

Pmax is an important measure of solar cell performance and is often used to compare different solar cells or modules.

In addition, the current-voltage characteristic of a photovoltaic cell shows a maximum power point P_{MPP} (MPP stands for Maximal Power Point).

13.2- Power-voltage characteristic (P-U) :

The power-voltage characteristic shows the relationship between the power generated by the solar cell and the voltage at its terminals. It can be used to identify the solar cell's optimum operating point, which corresponds to the maximum power delivered by the cell. The power delivered by the cell is expressed as P = U.I. For each point on the above curve, we can calculate the power P and plot the curve P = f(U).

This curve looks like this:

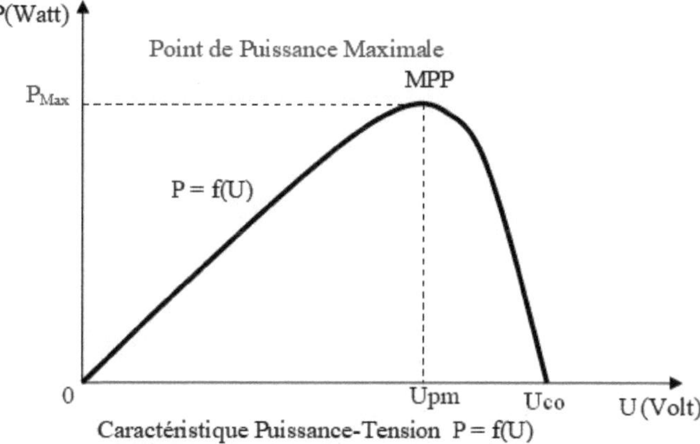

Figure 29: P=f(U) characteristic of the solar cell

This curve passes through a power maximum (Pmax).

This power corresponds to a voltage Upm and a current Ipm, which can also be seen on the curve I = f(U).

14- Influence of irradiance and temperature on the yield of a photovoltaic panel :

Efficiency depends on factors such as the quality of the materials used, cell design, solar illumination conditions and ambient temperature. Solar cell manufacturers generally provide detailed data in their product data sheets.

As mentioned above, the efficiency of photovoltaic panels varies according to the technology used to manufacture the solar cells, but there are other factors that influence their performance. Before discussing these factors, it's important to understand what STC (Standard Test Conditions) and NOCT (Normal Operating Cell Temperature) conditions are.

STC (Standard Test Conditions) and NOCT (Normal Operating Cell Temperature) are two sets of standard conditions used to characterize solar cell performance. Here are the main differences between STC and NOCT:

Measurement conditions :

-STC (Standard Test Conditions): STC measurements are carried out under ideal laboratory conditions, including an irradiance of 1000 W/m², a cell temperature of 25°C, and an air mass spectrum (AM) of 1.5.

-NOCT (Normal Operating Cell Temperature): NOCT measurements simulate actual solar cell operating conditions. They include an ambient temperature of 20°C, an irradiance of 800 W/m², and a wind speed of 1 m/s. These conditions are closer to those to which a solar cell would be exposed under normal operating conditions.

14.1-Influence of illumination :

Current-voltage characteristic I = f(U) of modules at different irradiances and constant temperature :

This curve shows that short-circuit current increases with illuminance, while open-circuit voltage varies little. At constant temperature, the I = f(U) characteristic is highly dependent on illuminance.

Variations in irradiation have the greatest influence on module current, as it depends directly on irradiation intensity. When irradiation is halved, the current produced is also halved.

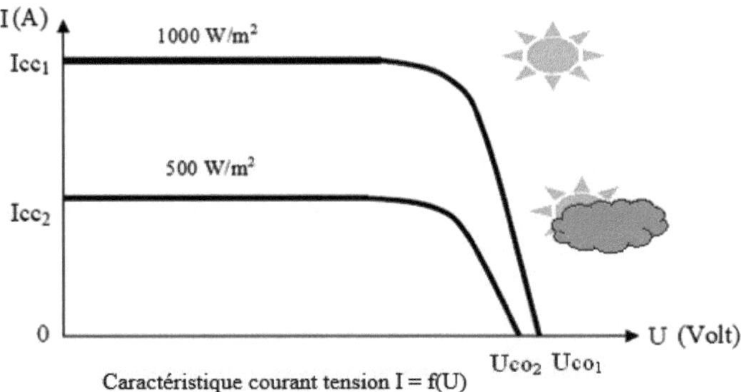

Figure 30: Influence of illuminance on I = f(U) characteristics

From these curves, we can draw the power curves P = f(U) :

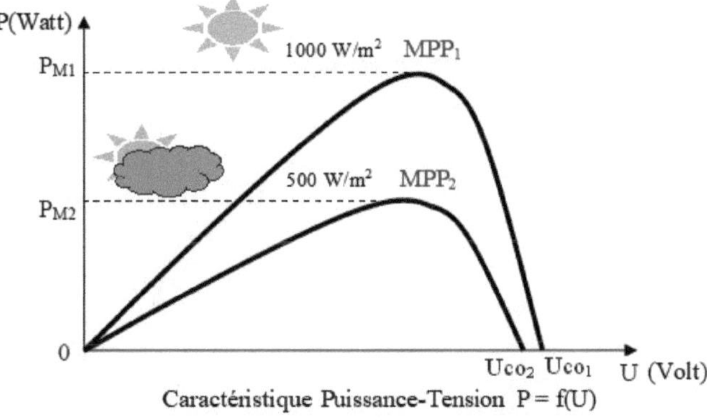

Figure 31: Influence of illuminance on P = f(U) characteristics

These curves show that the maximum power delivered by the cell increases with illuminance.

14.2-Temperature influence :

Current-voltage characteristic of modules at different temperatures and constant irradiation (1000 W/m²):

Module temperature has the greatest influence on voltage. For this reason, it is particularly important to take into account the rise in voltage at low temperatures.

For a fixed illuminance, the characteristics I = f(U) and P = f(U) vary with the temperature of the photovoltaic cell:

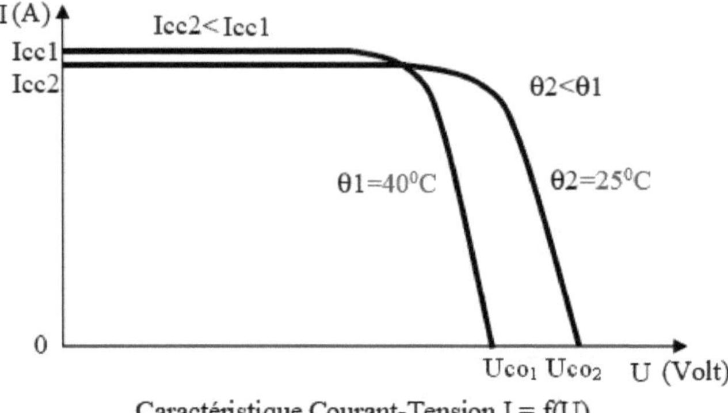

Figure 32: Influence of temperature on I = f(U) characteristics

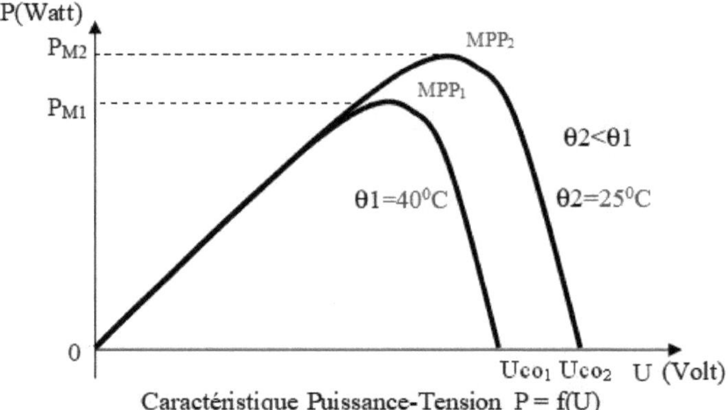

Figure 33: Influence of temperature on P = f(U) characteristics

These curves show that no-load voltage and maximum power decrease with increasing temperature.

When: T increases we see that: $Uco_2 > Uco_1$ and $Pm_2 > Pm_1$ while $Icc_2 < Icc_1$.

Note: The voltage rise at low temperatures must be taken into account.

14.3-Influence of Positioning : The energy supplied by the PV panel is highly dependent on the amount of solar irradiance absorbed by the panel.

This amount depends on the panel's orientation in relation to the sun. To collect the maximum amount of energy, the PV panel must be constantly oriented perpendicular to the sun's rays. The optimum angle is 90°.

14.4-Influence of the tilt angle: Another factor influencing PV panel performance is the tilt angle, which corresponds to the angle formed by the plane of the solar panel in relation to the horizontal (the ground plane). The tilt angle is smaller in summer and larger in winter.

Note: Factors such as shading, panel cleanliness, etc., can also influence the actual performance of a photovoltaic system.

Definition of peak power :

The peak power (**Pc**) of a photovoltaic panel is the maximum power delivered by a cell under specific conditions. These conditions include solar irradiance (incident sunlight) of **1000 W/m²**, a temperature of **25°C**, and a spectral distribution of radiation corresponding to the **AM 1.5** spectrum.

The unit of power is the Watt peak, or **Wp**.

Note: However, it's important to note that peak power is rarely achieved in real-life conditions. In practice, solar irradiance may be less than 1000 W/m², and the temperature of the solar panels may vary according to weather conditions.

15-Energy conversion efficiency (η) :

The energy conversion efficiency (η) of a photovoltaic system is a key indicator of its efficiency. It is defined as the ratio between the maximum power supplied by the solar module (Pmax) and the power of the incident solar radiation (Pinc).

$$\eta = \frac{Pmax}{Pinc} \qquad ; \eta = \frac{Umax * Imax}{S * G}$$

Where:

Pmax: is the maximum power supplied by the module (W).

Umax : is the module's maximum voltage (Volt).

Imax : is the maximum module current (A).

S : is the surface area of the solar cell in (m^2).

G : is illuminance in (W/m^2).

16-Form factor (*FF*) :

Another important parameter that indicates the degree of ideality is the so-called fill factor. It represents the quality of the solar cell and is defined as the ratio between the actual maximum power supplied by the cell (Pmax) and the product of the short-circuit current (Icc) and the open-circuit voltage (Uco), which represents the maximum power of a theoretically ideal cell.

$$FF = \frac{Pmax}{Uco*Icc} \qquad FF = \frac{Umax*Imax}{Uco*Icc}$$

The form factor generally varies around 0.7-0.8 for high-performance solar cells, and decreases with rising temperature. It gives an idea of the cell's quality. These parameters are crucial for assessing the overall efficiency and quality of a photovoltaic module.

17-Coefficient of performance Per :

The coefficient of performance **Per** to indicate the ratio between the actual observed efficiency $\eta_{réel}$ of a photovoltaic cell and the theoretical maximum expected efficiency η.

$$Per = \frac{\eta\ réel}{\eta\ théorique}$$

It's important to note that actual yields are often influenced by various factors, such as weather conditions, the angle of incidence of the sun, the quality of materials, etc. Theoretical yield is based on the intrinsic characteristics of the photovoltaic cell.

Example:

Electrical characteristics (at 1000 W/m)^2		
Cell temperature (°C)	25	50
Maximum power (Pmax in watts)	36	32.5
Voltage at maximum power (Umax in V)	16.3	14.4
Short-circuit current (I dc in A)	2.45	2.50
Open circuit voltage (U co in Volt)	20.3	18.4
Current at 10 V	2.29	2.28
Current at maximum power (Imax in A)	2.21	2.26
Efficiency η (For a 1 m² cell surface)	3.6%	3.25%
FF form factor	0.873	0.791
Per performance coefficient (for $_{theoretical}$ η =4%)	0.9	0.8125

Table 10: Electrical characteristics of PV at 1000 W/m^2

18-Different types of photovoltaic systems :

18.1-Off-grid photovoltaic systems: Independent of the public electricity grid, they use batteries to store the electricity produced, and are often used in isolated areas.

18.2-Grid-tied photovoltaic systems: Connected to the public electricity grid, they can sell the electricity they produce and buy it when needed.

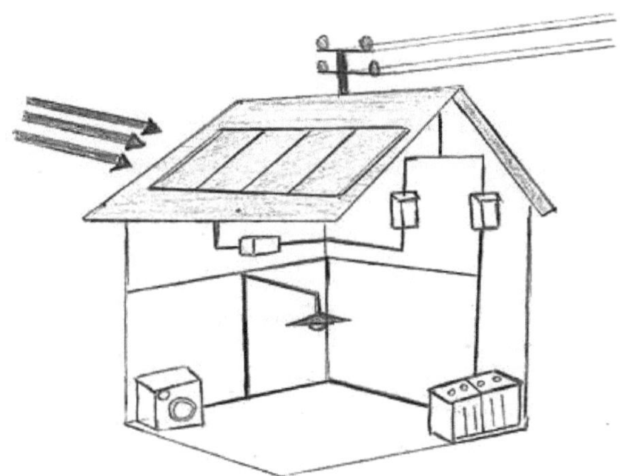

Figure 34: Grid-tied PV systems

19-Structures of a photovoltaic system :

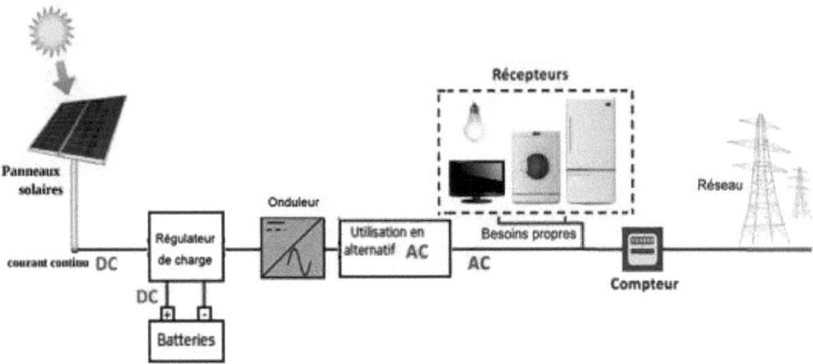

Figure 35: A **photovoltaic** system

20-Protection of PV modules :

In photovoltaics, the term **"hot spot"** occurs when a solar panel cell is defective or partially shaded, creating high electrical resistance. This prevents the cell from generating electricity efficiently and causes a build-up of heat, which can damage the panel, shorten its lifespan and, in extreme cases, lead to fire.

To prevent "hot spot" problems, solar panel designers often integrate **"by-pass" diodes** into the panels. These diodes bypass the current around defective or shaded cells, reducing the risk of overheating in these specific areas.

To prevent current flowing in the reverse direction of the string, **blocking diodes**, also known as **anti-reverse diodes**, are electronic components used in solar systems to prevent energy losses caused by the reverse current phenomenon.

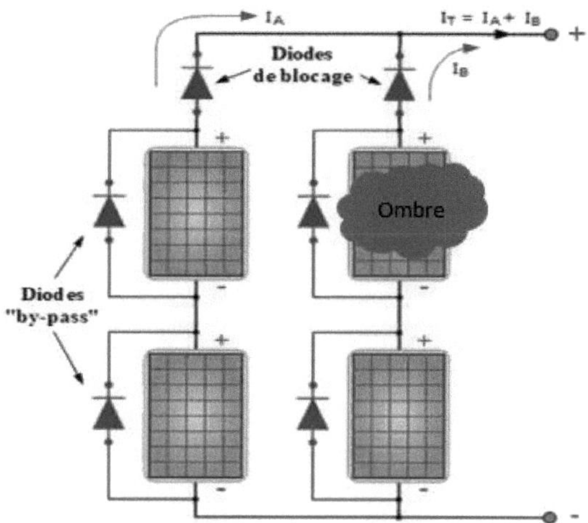

Figure 36: PV module protection

21-Maintenance and cleaning of photovoltaic panels :

21.1-Pollution from solar panels :

Soiling on an open-air surface (such as a solar panel) is made up of a mixture of organic and inorganic materials. Some of these materials will be suspended in the wind, while others will be deposited after evaporation (rain, water, thaw, fog).

Various solar energy studies show that solar panels lose between 5% and 16% of their efficiency due to the accumulation of dirt and dust on the panels and frame edges. Worldwide technology is focused on the direct and periodic cleaning of solar panels, but this method can cause damage if the procedure and products used are not respected. There are various ways of cleaning solar panels, such as :

Manual cleaning: manual brushing.

Semi-automatic cleaning: where the human controls the cleaning process.

Automatic cleaning: detects dust and carries out the cleaning process automatically, without human intervention, e.g. robotic scrubbers, drones and robots.

22-Advantages and disadvantages of photovoltaic energy :

Photovoltaic technology offers a number of advantages:

-Photovoltaic energy is free renewable energy.

-Photovoltaic systems can be installed anywhere, even in cities.

-Photovoltaic energy offers a practical solution for obtaining electricity in isolated towns.

-Photovoltaic electricity is generated as close as possible to its point of consumption, in a decentralized way, directly at the user's premises.

-Photovoltaic systems are extremely reliable.

-Photovoltaic panels have a very long lifespan.

However, there are also disadvantages to photovoltaic energy:

The manufacture of photovoltaic modules is a high-tech, capital-intensive process.

-The actual conversion efficiency of a module is low.

-They are very sensitive.

-Maintenance costs such as cleaning and security for photovoltaic solar panels are relatively low.

Passive solar energy

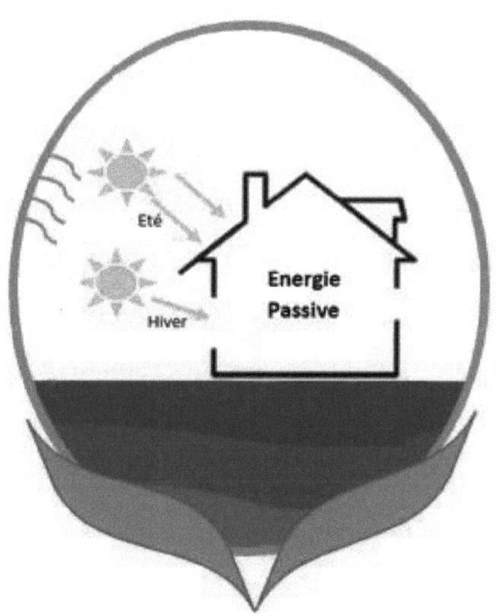

1-Passive solar energy basics :

Passive solar energy, a solution not to be overlooked!

Passive energy refers to the use of specific design and construction techniques in buildings to minimize energy consumption for heating, cooling and lighting. Unlike active systems that require mechanical equipment such as pumps or fans, passive energy relies on simple architectural principles to take advantage of natural conditions such as sunlight, natural ventilation and thermal insulation.

***A system that collects solar energy directly through the windows**:

A system that captures the sun's rays through the home's windows. The diagram opposite shows the variation in solar radiation intensity through differently oriented windows on a winter or summer day. Clearly, windows play a fundamental role in the thermal comfort of a home. Depending on the orientation of the windows, solar radiation will be more or less intense.

***An indirect solar energy collection system using a collector wall:**

A system that captures solar energy by means of a collector wall. This is made of heavy materials and is located between the space to be heated and the glass that protects it from the outside. The heat thus generated on the outside during the period of exposure to the sun is stored by the mass of the wall. It is then transferred to the interior volume. From there, it spreads to the compartment by convection and radiation.

2-A passive solar house :

Figure 37: Passive solar house

A passive solar home is designed to dispense with heating in winter and air conditioning in summer, while offering optimal thermal comfort. Thanks to its structure, design, orientation, the quality of its materials and the performance of its ventilation, it captures solar radiation to heat the house in winter, and shades it in summer to prevent overheating. Although self-sufficient in heating, it depends on the electrical grid for other needs. For energy self-sufficiency, solar panels can be added to the roof to power electrical appliances and cover lighting needs.

3-Passive solar house operation :

A passive solar home heats solely with thermal energy from the sun, occupants (humans and animals) and electrical appliances. To achieve this, it must maximize sunlight and eliminate heat loss by being well exposed, with optimal openings, built with quality materials, and featuring a coherent interior and exterior layout. For example, trees that shed their leaves in winter to let in the sun and create shade in summer.

4-Passive building design highlights :

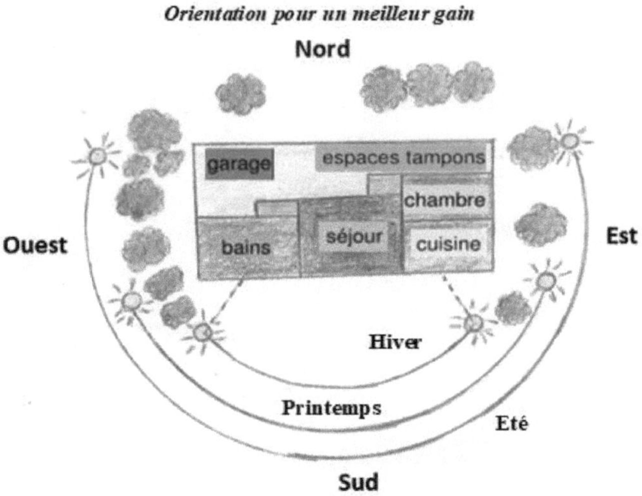

Figure 38: **Passive** building design

***Room layout :**

-The living rooms (lounge, dining room) face south.

-Rooms face east or west.

-Technical or cold rooms (garage, storeroom) are located to the north.

***Openings** :

The windows and French windows are arranged on each wall of the house:

*50% to the south (bay windows),

*20% to the west (small openings),

*20% to the east (small openings),

*10% to the north (small openings).

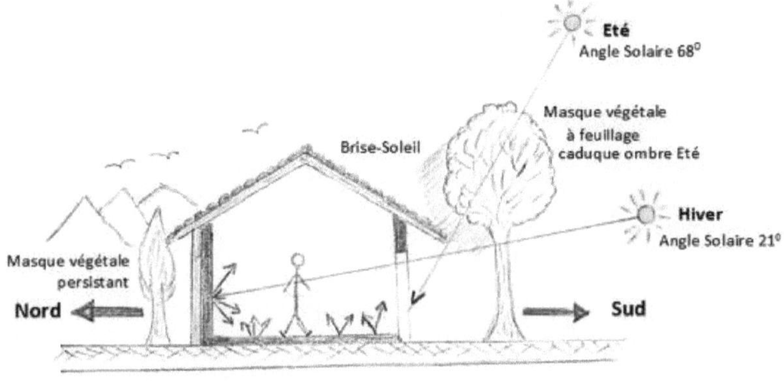

Figure 39: Orientation and layout of a passive solar house

For a passive solar house, here are the basic principles:

***Orientation and layout** :

-Solar orientation: Maximize solar gain in winter by orienting large windows towards the south for the northern hemisphere, but the opposite for the southern hemisphere.

***Part design :**

-Living areas: Rooms where occupants spend the most time (such as the living room, kitchen and dining room) should face south (or north in the southern hemisphere).

-Utility rooms: Rooms requiring less heating, such as bathrooms, laundry rooms and garages, can be placed on the north side (or south in the southern hemisphere) to act as buffer zones.

***Windows and openings :**

South-facing windows: Use large south-facing windows to maximize solar gain in winter.

-Windows to the north: Smaller or minimized windows to the north to minimize heat loss.

-East and west windows: Use moderate windows with shading devices to avoid overheating in the morning and afternoon.

***Insulation and thermal inertia** :

Insulation: Use of high-performance insulation to minimize heat loss in winter and heat gain in summer.

-Thermal inertia: choosing materials such as stone or concrete that can store heat during the day and release it at night. The choice of materials is essential to achieve high-quality insulation, avoid heat loss and ensure comfort in summer. All thermal bridges must be eliminated.

***Natural ventilation**:

Cross-ventilation: Design that encourages air circulation to cool the house naturally.

-Night ventilation: Use of day/night temperature differences to cool the house.

***Sun protection** :

-Brise-soleil and awnings: Devices to protect windows from direct sunlight in summer, while letting in light in winter.

-Vegetation: Plant deciduous trees to the south to provide shade in summer and allow light to pass through in winter, such as **Vines (**In addition to grapes, some ornamental deciduous vines can be used to cover structures in summer and lose their leaves in winter). And plant evergreen trees to the north to protect the habitat from winter winds.

5-Benefits of passive solar homes :

***Energy savings**: consume 7 to 8 times less energy for heating than a conventional house, with even greater savings if solar panels are used.

***Thermal comfort**: thanks to their high thermal inertia, they offer a gentle, even temperature with slow temperature variations.

***Sustainability**: energy performance remains constant over the years, unlike traditional heating systems.

***Low carbon footprint**: Reduced carbon emissions, especially with the use of renewable energies and bio-sourced materials.

***Resale appeal**: Compliant with strict energy standards and energy-efficient, they are highly sought-after on the real estate market.

6-Disadvantages of passive solar homes :

***High cost**: Construction is more expensive due to design requirements, high-quality materials and specialized equipment.

***Constraining design**: Architectural constraints limit design freedom, which can be frustrating for homeowners.

***Difficult location**: Requires clear, well-oriented ground, which can be problematic in urban areas or natural environments.

***Summer comfort**: Overheating can occur during the warmer months, even with solar protection devices.

7-Conclusion :

Thanks to its innovative design, a passive solar home offers efficient heating in winter without the need for conventional heating systems. Optimum orientation, strategic placement of openings, well thought-out interior layouts and the use of high-quality materials enable these homes to capture and retain solar heat. What's more, the addition of solar panels on the roof can transform this home into an almost entirely self-sufficient one, reducing your ecological footprint and energy costs. A passive solar home is therefore a wise investment in a comfortable, sustainable future. As you've probably gathered by now, passive solar heating offers many advantages for new-build properties. Both modern and environmentally friendly, it saves energy and helps you manage your heating bill more effectively. Why not consider it for your construction project?

Solar thermal energy

1-Definition :

Solar thermal energy refers to the transformation of solar energy into heat, used primarily for domestic and industrial heating and domestic hot water production. It is a renewable form of energy that helps reduce dependence on fossil fuels and the carbon footprint of heating systems.

2-Solar thermal panel :

2.1-Definition :

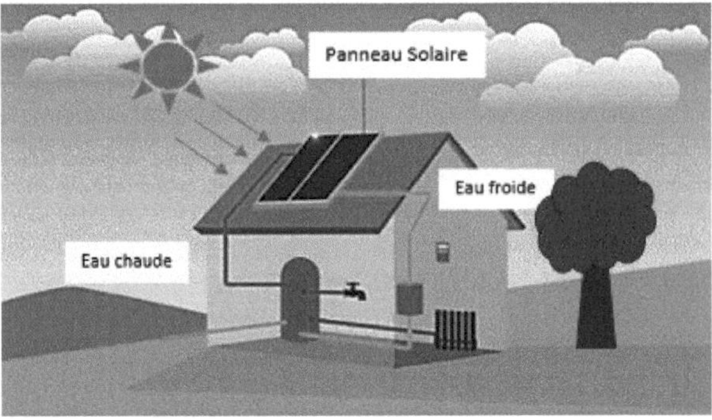

Figure 40: House with solar thermal panel

Taking the form of a rectangular module installed on your roof, like other types of solar panels (photovoltaic), thermal panels use solar energy to meet the energy needs of your home.
While photovoltaic panels use sunlight to generate electricity, thermal panels harness the sun's heat to heat water in your home, ecologically and free of charge. Thermal solar panels can therefore supply your home with domestic hot water, as well as your heating system if it's connected to a hot water circuit (underfloor heating or water radiators).

2.2-Solar thermal panel operation :

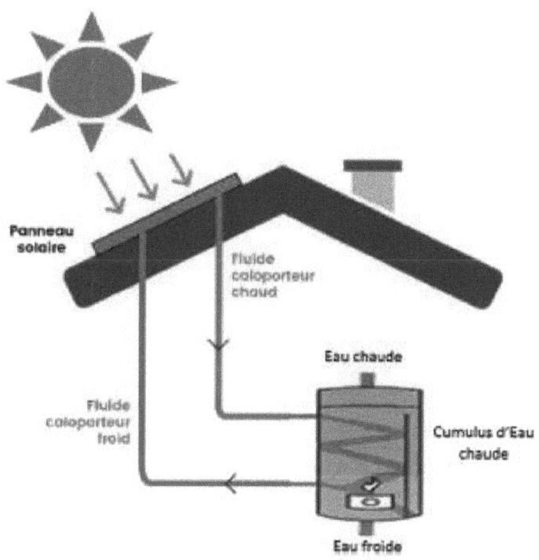

Figure 41: How a solar **thermal** panel works

A solar thermal panel is a device that converts the energy of sunlight into thermal energy (heat). The thermal energy is then absorbed by a heat-transfer fluid, such as water or air. The fluid circulates in a black-painted coil, which may be covered by a glass surface and protected on the other sides by insulation. Water-source thermal panels can be used to heat domestic hot water or as auxiliary heating. Solar thermal energy can also be used to heat a house via underfloor heating. In this case, heat is transferred directly to the air. When there isn't enough sunshine to bring the water up to the right temperature, the water heater can also be connected to a boiler to supplement the panel.

3-Construction of a solar thermal panel :

A solar thermal panel is made up of collectors that harness the sun's heat. This raises the temperature of a heat-transfer fluid. This fluid is then used to heat :

Water from the heating circuit to your radiators.

The water in a hot water tank.

When the sun fails to shine, a system such as an electric resistance or a boiler takes over to ensure continuous heat production.

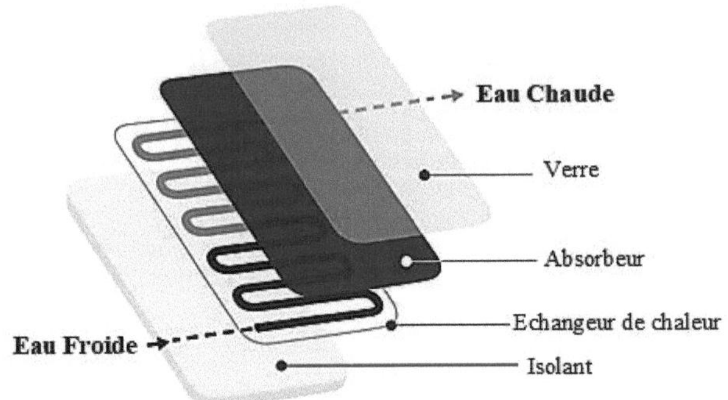

Figure 42: Construction of a solar **thermal** panel

4-Thermal applications of solar energy :

Although the use of solar energy has intensified in recent years, and technologies are constantly evolving, the principles and phenomena involved are very well known, and are now the subject of broad consensus. Solar radiation has been the subject of a great deal of study, research and discussion, ranging from low-temperature applications (water heating, home air-conditioning, water desalination, solar refrigeration, etc.) to relatively higher-temperature applications (solar cookers and ovens), which require the radiation to be concentrated.

Without being exhaustive, we can mention :

***Solar hot water :**

Domestic hot water production is currently the most widespread application of solar thermal energy. The classic installation essentially comprises a group of flat-plate collectors, a storage capacity and, generally, a control system and a back-up source. The temperature of domestic hot water is relatively low, and flat-plate solar collectors are well suited to this type of hot water production.

***Solar home heating:**

Solar heating of buildings uses air or water collectors, but with larger surface areas per dwelling. Heat can be distributed via hot-water radiators or underfloor or ceiling heating. The use of solar heat for building heating requires some form of storage. It is possible to store energy in the form of hot water in tanks of several cubic meters in size, thus overcoming the intermittent nature of solar energy.

5-The different types of thermal panels :

Thermal panels all work in the same way, but they're not all manufactured in the same way. There are 3 main types of solar thermal panels.

5.1-Opaque flat-plate collectors :

These solar thermal panels consist of opaque, dark collectors in which a heat-transfer liquid is circulated. The sun strikes the collectors directly. The fluid then rises in temperature.

These collectors are not the most efficient. Although they heat up quickly, they let a lot of surface heat escape because they are not insulated with a surface layer, unlike flat-plate glass collectors, for example.

On the other hand, because they're easier to manufacture and lighter, they're often more economical to buy. For example, they can be used to heat a swimming pool at lower cost.

5.2-Glazed flat-plate collectors :

Glazed flat-plate solar thermal panels are the most widespread. These are dark solar collectors through which a heat-transfer liquid flows, and which are protected by insulation. A pane of glass is placed above the collectors to create a greenhouse effect and increase heat production.

5.3-Vacuum tube collectors :

These thermal panels are made up of a set of tubes in which a heat-transfer fluid circulates. Thanks to the air gap, these tubes are highly insulated. As a result, they lose very little heat at the surface. This technology is the most efficient, but also the most expensive.

6-Benefits :

Ecology: they produce non-polluting, 100% renewable energy

Savings: they considerably reduce your electricity bills

Self-consumption: they help you gain autonomy and energy independence

Resistance and durability: they last a long time (between 25 and 30 years)

Practical: low-maintenance

Real estate value: they add value to your home.

7-Disadvantages :

High investment: relatively high purchase and installation costs

Low efficiency in winter: this will most often require the installation of auxiliary heating.

8-Calculation of instantaneous yield η :

The efficiency of a solar collector, whose symbol is η, is the ratio of the heat stored by the heat transfer fluid Q to the incident power received by solar radiation G .$_i$

$$\eta = Q / G_i$$

G_i : Global irradiance incident on the collector (**W/m^2**).

Thermodynamic solar energy

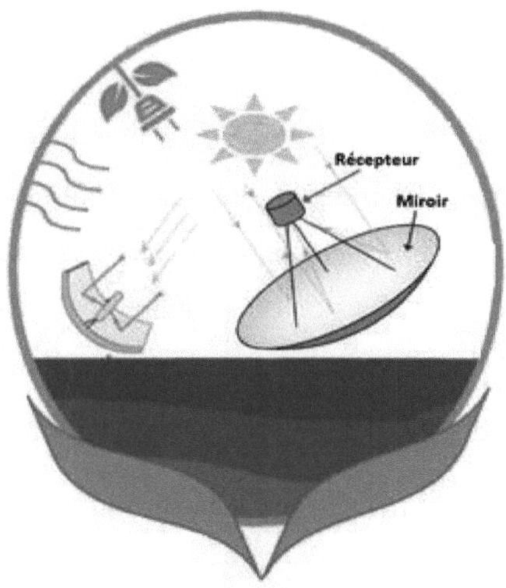

1-Definition :

Thermodynamic solar power is an advanced renewable energy technology that uses concentrated solar radiation to generate electricity from heat. Unlike photovoltaic panels, which convert sunlight directly into electricity, thermodynamic solar harnesses the intense heat generated by solar radiation. This concentrated thermal system achieves much higher temperature levels than conventional, non-concentrated thermal systems. Whereas domestic water heaters produce water at around 50 degrees Celsius, it is possible, through concentration, to heat fluids to temperatures of the order of 250 to 1000°C. The heat is then converted into mechanical energy, then into electricity via turbines, similar to those used in traditional power plants. So photovoltaics is not the only solar energy capable of generating electricity. A key feature of this technology is the ability to store heat in thermal reservoirs, enabling continuous electricity production even in the absence of sunlight. It therefore represents an efficient and sustainable solution for electricity generation, capable of meeting energy needs while minimizing environmental impact.

2-Principle of concentration (A little history) :

Historically, the first solar concentration systems date back to antiquity, when the Greeks and Romans used mirrors to concentrate sunlight, as illustrated by the myth of Archimedes' "burning mirrors". Most often using reflecting mirrors or magnifying glasses, a concentrating system redirects the solar radiation collected by a given surface onto a smaller target: starting a fire of fallen leaves with a magnifying glass uses this principle. However, the systematic use of solar concentration to generate heat and produce energy didn't really take off until the 19th century. At that time, Augustin Mouchot, a French inventor, developed devices using parabolic mirrors to concentrate sunlight and produce steam, used to power engines and for industrial purposes (Solar furnace by Augustin Mouchot and Abel Pifre circa 1880).Over the years, this technology has been perfected and adapted to different applications, such as electricity generation from Concentrated Solar Power (CSP) plants, where mirrors or lenses are used to concentrate sunlight on a receiver that converts the heat into electricity. Today, the principle of concentrated solar power continues to be explored and developed to improve energy efficiency and reduce costs, contributing to the transition to more sustainable renewable energy sources.

3-Solar thermodynamic operating principle :

Thermodynamic solar power plants use a large number of mirrors to converge the sun's rays onto a high-temperature heat transfer fluid. To achieve this, the reflective mirrors must follow the sun's movement to capture and concentrate radiation throughout the daily solar cycle. The fluid then generates electricity via steam or gas turbines. Here are the main steps in the process:

-Concentration of solar radiation: mirrors or lenses, often arranged in fields, concentrate sunlight on a focal point or line. These systems are designed to maximize the capture of solar energy, by directing the radiation towards a receiver.

-Fluid heating: The receiver, placed at the focal point, captures this concentrated energy and transfers the heat to a heat transfer fluid (such as oil, molten salts, or water). This fluid is heated to temperatures of up to several hundred degrees Celsius.

-Converting heat into electricity: The hot fluid is then used to produce steam in a heat exchanger. The steam generated turns a turbine, which is connected to an electricity generator. The process is similar to that of conventional thermal power plants, except that the heat source is the sun.

-Heat storage: Some of the heat can be stored in thermal tanks for later use. This thermal storage enables electricity generation to continue even when the sun is not available, for example at night or on cloudy days.

-Electricity distribution: The electricity generated is then fed into the power grid for distribution to consumers.

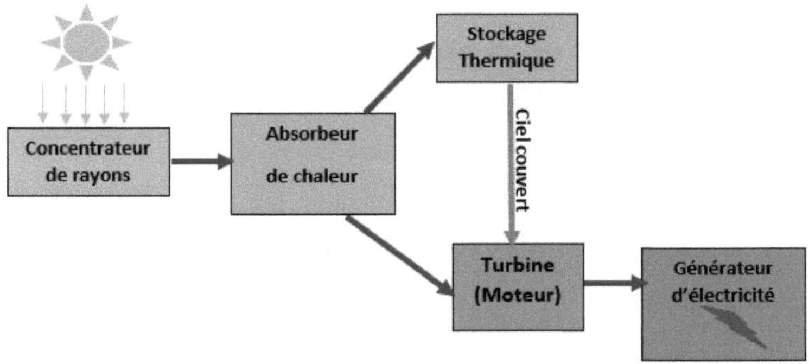

Figure 43: Solar concentrator operating principle

4-The main types of power plants :

The CSP (Concentrated Solar Power) system generates electricity by converting solar energy into high-temperature heat using reflectors and receivers. The heat is then used to generate electricity through a conventional turbine-generator system. Large-scale CSP plants can be equipped with a heat storage system to enable heat supply or electricity generation at night or on cloudy days.

There are two types of solar concentrators:

-Linear concentrators :

Concentration takes place on long tubes in which a heat transfer fluid circulates. These tubes are located on the focal line of the reflectors concentrating the solar radiation. This technology requires the sun to be tracked on one axis. Parabolic trough and Fresnel collectors work on this principle.

-Point concentrators :

Concentration takes place on a central receiver. The concentrator follows the sun on two axes: azimuth and elevation. This principle is used by parabolic concentrators and tower power plants.

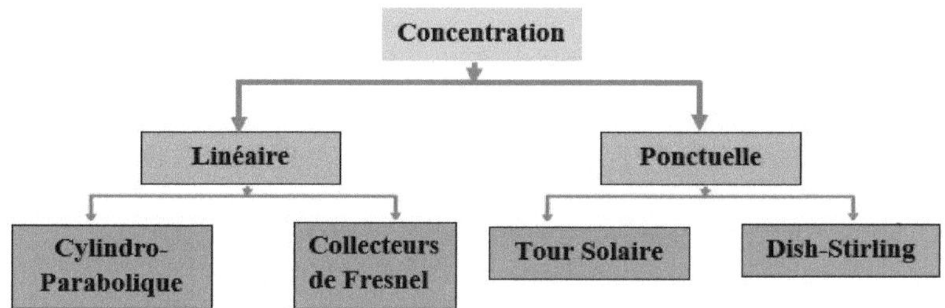

Figure 44: Classification of different types of solar concentrators

4.1-Linear concentration systems :

Solar radiation is concentrated on one or more absorber tubes installed along the focal line of the mirrors. This tube contains a heat-transfer fluid heated to between 250 and 500°C. The reflective mirrors follow the Sun's movement throughout the day.

a-Parabolic trough power plants :

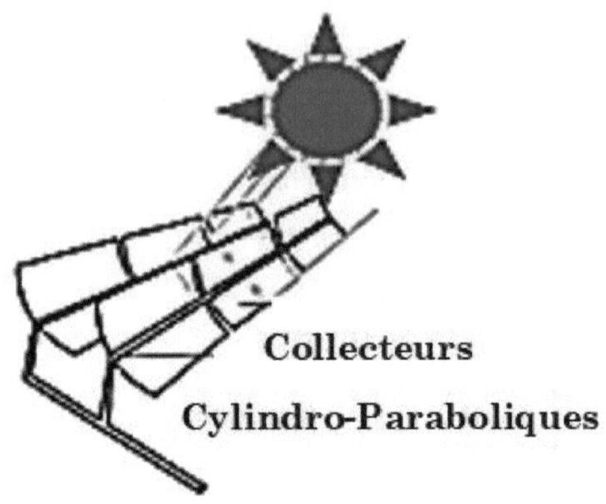

Figure 45: Parabolic trough sensors (Linear Concentration, Mobile).

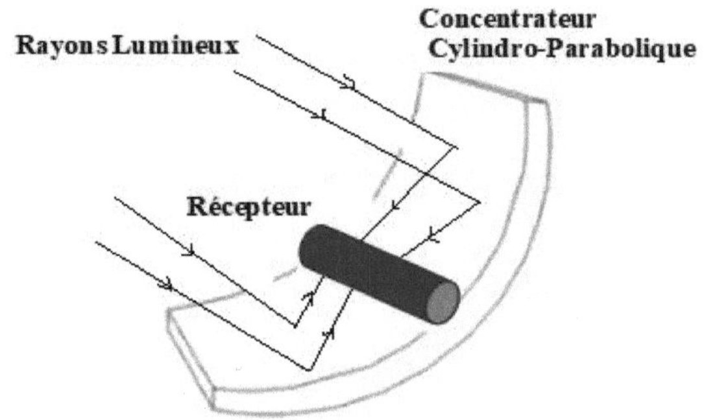

Figure 46: Optical system of a Cylindro-Parabolic concentrator

Parabolic trough collectors are made up of curved mirrors in the shape of an elongated parabola (cylindrical), which concentrate sunlight along a focal line. This focal line is parallel to the axis of the mirrors. Sunlight is concentrated on an absorber tube running along the focal line of the mirror. This tube contains a heat transfer fluid (such as thermal oil, water, molten salts or air) which is heated by the concentrated light. These power plants generally reach temperatures of between 200 and 500°C. With efficiencies of up to 10-14%. They are commonly used for large-scale applications, such as power generation and the supply of process heat at moderate temperatures.

b-Fresnel linear sensor power plants :

Figure 47: Linear Fresnel sensors (Linear Concentration, Fixed).

They use flat mirrors arranged in parabolic rows to concentrate sunlight on receiver tubes located above the rows of mirrors. These tubes contain a heat-transfer fluid (such as water, molten salts or oil). Less intense than tower power plants, but still capable of efficiently concentrating sunlight. They can reach temperatures of 200°C - 450°C, generally sufficient to power steam turbines. Mainly used to generate electricity by heating a heat-transfer fluid, which is then fed into steam turbines.

4.2-Point concentration systems :

The sun's rays are concentrated around 1,000 times in a single, small focus. The temperature can reach 500 to 1000°C.

a-Dish-Stirling parabolic trough power plants :

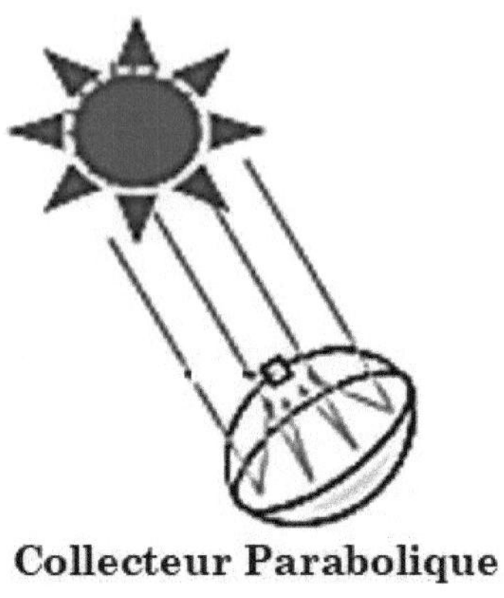

Figure 48: Dish-Stirling parabolic sensors (Punctual Concentration, Mobile).

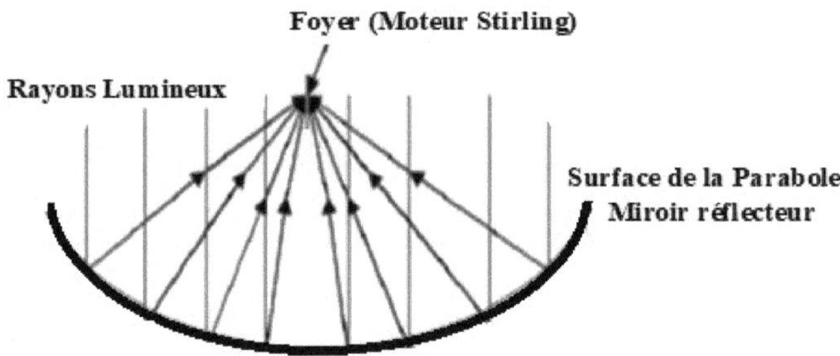

Figure 49: Optical system of a parabolic concentrator

Dish-Stirling parabolic trough power plants are distinguished by their innovative design. Unlike parabolic trough collectors, parabolic trough collectors use parabolic mirrors to concentrate sunlight in a single focal point, located above the mirror, to drive a so-called "Dish-Stirling" motor. Once heated in a closed circuit, the gas it contains activates a piston that recovers the mechanical energy produced. This intense concentration enables extremely high temperatures to be reached, up to 300°C - 1000°C. Dish-Stirling systems also stand out for their remarkable efficiency in converting solar thermal energy into electricity, with efficiencies of up to

20% to 31% in the ten kWe range. Unlike other technologies such as solar towers and parabolic trough concentrators, which are optimized for megawatts or more, Dish-Stirling power plants are well suited to modest-sized installations and can operate autonomously, ideal for isolated sites not connected to the power grid.

b-Tower power plants (Solar tower) :

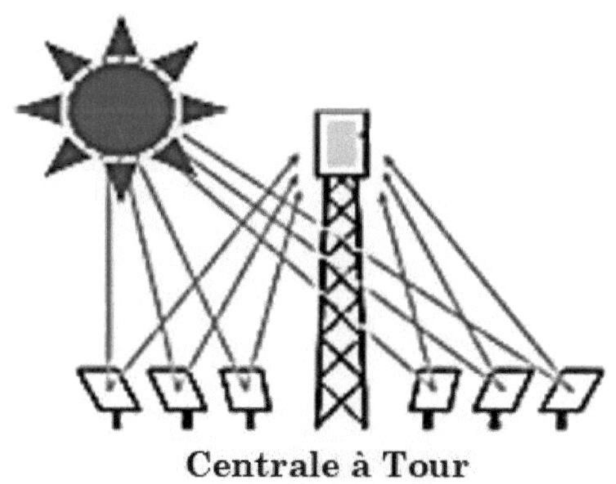

Figure 50: Solar tower (Concentration Punctual, Fixed)

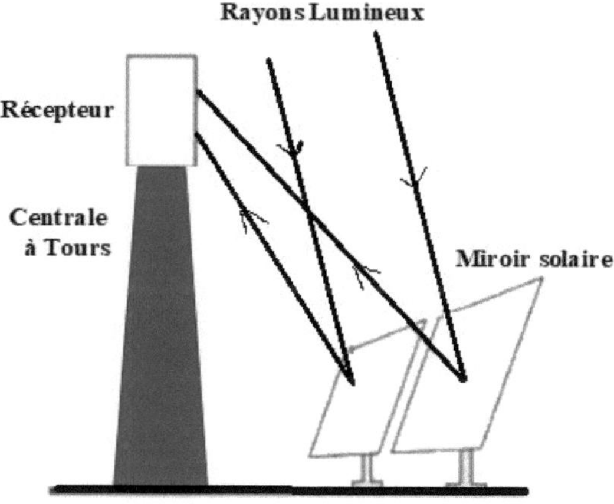

Figure 51: Optical system of a solar tower concentrator

These power plants use mirrors or reflectors to concentrate sunlight towards a central point, where a tower is located. The mirrors are often arranged in a circle around the tower. Sunlight is significantly concentrated to reach high temperatures at the top of the tower. These can reach a thousand degrees Celsius, 300°C - 1000°C, depending on the precision of the solar concentration. Solar-electricity efficiency of 12-15%. Mainly used to generate electricity by converting concentrated heat into electrical energy using turbines or Stirling engines.

5-Benefits of thermodynamic solar energy :

Renewable and clean: uses solar energy, an inexhaustible, non-polluting resource.

-Thermal storage: Heat can be stored to generate electricity even when the sun is not shining.

-Energy stability: Can complement intermittent renewable energies such as wind power and photovoltaics.

-Long service life: The plants have a long service life and can operate for several decades with proper maintenance.

6-Disadvantages of solar thermodynamic energy :

-High cost: Initial construction and maintenance costs are high compared to other renewable technologies.

-Environmental impact: Requires large areas of land, which can disrupt local ecosystems.

-Weather dependency: Performance is strongly influenced by weather conditions, requiring direct sunlight.

-Technical complexity: Thermodynamic power plants are technically complex, requiring specialized expertise for their design, construction and maintenance.

References :

[1] Jacques Bernard, Energie Solaire, calculs et optimisation, Ellipse Edition Marketing ,9782729864927, (2004).

[2] Cid Pastor Angel: "Conception Et Réalisation De Module Photovoltaïques Electroniques", l'institut national des sciences appliquées de Toulouse, 2006.

[3] A. Goetzberger, V. U. Hoffmann, Photovoltaic Solar Energy Generation, Springer Series in Optical Sciences, Atlanta (2005).

[4] J.M. Chassériau, Conversion thermique du rayonnement solaire, Bordas, Paris (1984).

[5] D.K. Edwards, Capteurs solaires, Edition SCM, Paris (1979).

[6] J.R. Vaillant, Utilisation et promesses de l'énergie solaire, Edition Eyrolle, 1976.

[7] J. Bonal, P. Rossetti, Alternative energies, Omniscience, 2007.

[8] Ch. Ngo, L'énergie ressources technologies et environnement, 3rd edition, Dunod, Paris, 2008.

[9] Alain Bilbao Learreta, "Réalisation de techniques MPPT numériques", Rapport de stage de fin d'études, https://www.hisse-et-oh.com/, Diplôme d'Ingénieur Technique Industrielle, Université Virgile, September 2006.

[10] A. Sfeir, G. Guarracino, Ingénierie des systèmes solaires, Technique et Documentation, Paris, 1981.

[11] Walker, G. (1980). Stirling Engines. Oxford.

[12] Spencer, J.W. 1971. Fourier series representation of the position of the sun. Search 2: 172.

[13] Solar Electricity: An Economic Approach to Solar Energy by W. Palz, Butterworths, UNESCO, 1978.

[14] Le Rayonnement Solaire by R. Bernard, G. Menguy, M. Schwartz, Technique et Documentation, Lavoisier, Paris, 1979.

[15] R. Bernard, G. Menguy, M. Schwartz, Le rayonnement solaire conversion thermique et applications, Technique et documentation Lavoisier, 2nd edition 1980.

[16] A. Sfeir, G. Guarracino, Ingénierie des Systèmes Solaire, application à l'habitat, Technique et Documentation, Paris, 1981.

[17] L'Énergie Solaire dans le Bâtiment by Ch. Chauliaguet, P. Baratçabal, and J. P. Bat-ellier, Eyrolles, Paris, 1981.

[18] Les Photopiles Solaires : Du Matériau au Dispositif ; Du Dispositif aux Applications by A. Laugier and J. A. Roger, Technique et Documentation, Lavoisier, Paris, 1981.

[19] Michel Capderou, Atlas Solaire de l'Algérie, Tome 1, 2 ; O.P.U., 1986.

[20] M. Iqbal, An Introduction to Solar Radiation, Academic Press, Canada, 1983.

[21] J.M Chassériau, Conversion thermique du rayonnement solaire, Dunod, 1984.

[22] Énergie Solaire Photovoltaïque, le Manuel du Professionnel by Anne Labouret - Michel Villoz.

[23] D.Yogi. Goswami, Principles of Solar Engineering, Third Edition. Advan kt.com/principles of solar engi.pdf.

[24] Commons.wikimedia.org/wiki/File.

[25] Dr. SALMI Mohamed Support de cours : Gisement solaire.

[26] Bernard Thonon and Philippe Malbranche, Journée de formation Séminaire " Questions de Sciences " CEA.fr, Questions de physique sur l'énergie solaire, 06/06/2012.

[27] Christine Blondel_college_France_presentation.pdf, https://www.college-de-france.fr.

[28] Le Rayonnement Thermique, Bilan Radiatif et Effet de Serre, by V. Daniel, 2003.

[29] N. BAKI, Solar Radiation from the Sun to the Earth, 978-620-69417-5; EUE, European University Publishing.

[30] N. BAKI, Solar Radiation from Sun to Earth, Carpe Solem, 978-620-6-78245-2; LAP LAMBERT Academic Publishing.

Corrected exercises

Exercise 1:

Which of these greenhouse gases are among the main contributors to global warming?

a) Hydrogen (H)$_2$

b) Ozone (O)$_3$

c) Methane (CH)$_4$

d) Carbon dioxide (CO)$_2$

Solutions :

c ,d

Exercise 2:

1-Calculate the solar height and azimuth when it's TSV (10am) on April 18 in Toulouse.

The latitude is 43.6°North. Day number is n= 108.

2- Calculate the true solar time TSV when legal time is TL (10h) on April 18 in Toulouse. Longitude is -1.37°East. Day number is n= 108. The time zone correction is C_1 =+1h; and for France summer time C =+1h.$_2$

Solutions :

1-Hence, the hour angle $\omega = 15 \times (TSV - 12) = -30°$.

The declination is calculated as follows:

Declination $\delta = 23.45 \times \sin(360 \times (n + 284) / 365) = 10.5°$.

Solar height h :

sin (h) = sin(θ) × sin(□) + cos(θ) × cos(□) × cos(□) = 0.742, Solar height h = 47.9 °.

Azimuth a:

sin (a) = cos(□) × sin(□) / cos(h) = -0.7337, azimuth a = -47.2°. (a= -47.2°East).

2-Through calculation, we obtain: Equation of Time Et = 0.7 minutes=0.01h.

We deduce: Solar hour $TSV=TL-C_1 -C_2 - \varphi /15+Et$

TSV = 10 - 1 -1+1.37°/15+ 0.7/60 = 8+0.09+0.01=8.10 hours=8h 6mn.

Exercise 3:

What is the light-to-electricity conversion efficiency of a photovoltaic solar panel?

a) 2 to 7% of

b) 60 to 80% of the

c) 35% to 50%.

d) 15 to 25%.

Solutions :

d

Exercise 4:

Perovskites are special mineralogical structures with the conductive properties of a semiconductor. The cost of manufacturing perovskite-based photovoltaic cells is much lower than that of silicon cells. Recent research shows that it is possible to achieve efficiencies close to 25% with photovoltaic cells using perovskites.

1-Briefly **explain** the operating principle of a semiconductor.

2-Determine for which part of the solar spectrum perovskites absorb the most radiation.

3-Explain the benefits of using perovskites combined with silicon to make a photovoltaic cell.

Solutions :

1-A semiconductor will absorb energy from solar radiation. This will allow valence electrons to pass through the band gap into the conduction band. Current will flow and the material will become a conductor.

2-Perovskites absorb mainly in the visible part of terrestrial radiation between 450 nm and 800 nm.

3-Combining the two materials, the semiconductor absorbs more peak light energy (between 380nm and 1000 nm), boosting panel efficiency. Silicon absorbs IR light between 800 nm and 1100 nm, completing the absorption range of perovskites.

Exercise 5:

150 identical solar cells are available. Each cell has a surface area of 0.015 m^2 and the following characteristics under AM1 illumination of 1000 W/m^2 :

Short-circuit current I cc = 2 A, open-circuit voltage U co = 0.5 V, form factor FF = 70%.

1-Calculate the power of one of these cells and its efficiency.

2-Connect all 10 cells in parallel in a single block, then connect the 15 blocks in series to form a panel. Calculate the open-circuit voltage and short-circuit current of the resulting panel.

3-If the panel's form factor is 70%, what is the peak power drawn from this panel?

4-If the maximum voltage corresponding to the power is Umax = 2.5 V, calculate the maximum current Imax.

Solution:

1-Solar cell **power** :

$P_{cell=I}\ cc \times U\ co$, $P_{cell=2}\ A\times 0.5\ V=1\ W$

-Solar cell efficiency :

The efficiency η of a solar cell is calculated from the form factor (FF): $\eta=P_{cell*}$ P incident×100.

The power of the light incident on the surface of the solar cell.

$$P_{incident=1000}\ W/m^2 \times 0.015\ m^2 = 15\ W.$$

Let's calculate the yield:

$$\eta = 1\ \mathbf{W/15}\ W \times 100 = 6.67\%$$

2-Panel open-circuit **voltage** and short-circuit current :

By connecting 10 cells in parallel to form a block, and mounting 15 blocks in series to form a panel :

For a block of 10 cells in parallel :

Short-circuit current $Icc_{(block)}=10\times 2=20$ A. Open-circuit voltage $Uco_{(block)}=0.5$ V .

For the 15-block panel in series :

Open circuit voltage $Uco_{(panel)}=15\times 0.5=7.5$ V. Short-circuit current $Icc_{(panel)}=20$ A.

3-Panel peak **power**

$P_{crête}$ of the panel can be calculated using the form factor (FF):

$P_{crête} = U\ co_{(panel)} \times I\ cc_{(panel)} \times FF$. $P_{crête} = 7.5\ V \times 20\ A \times 0.7 = 105\ W$

4-Calculation of maximum current Imax :

If the maximum voltage corresponding to the power is Umax=2.5 V, the maximum current Im can be calculated by the relation :

$I\ max = P_{crête} \times U\ max$, $I\ max = 105\ W \times 2.5\ V = 42\ A$; $I\ max = 42\ A$.

Exercise 6:

Today, many buildings exploit renewable energies, in particular solar thermal energy. In some homes, hot water is obtained using solar thermal panels.

1- Diagram the energy transfers and conversions in a solar thermal panel.

2-The flow rate of the heat transfer fluid (water) circulating inside the pipes is $D=60\ L.h^{-1}$. The water enters at temperature Te= 20°C and leaves at temperature Ts=56°C. Calculate the energy received by the water over a period of one hour. Deduce the power supplied by the panel. Data: to raise the temperature of 1kg (i.e. 1liter) of water by 1°C, 5 kJ of energy must be supplied under the panel's operating conditions.

3-The solar thermal panel has a surface area of $S=3\ m^2$. Calculate the power received by this panel for a light output of $1000 W/m^2$.

4-Calculate the efficiency η of this panel.

Solution:

1-Energy conversion and transfer **diagram** for a solar thermal panel :

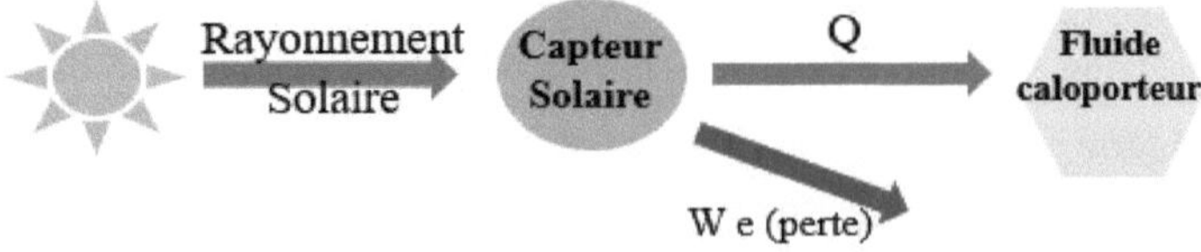

2- In 1 hour, 50 L of water circulate in the circuit under the panel, receiving the sun's energy by radiation.

The temperature rise of the water is: $\Delta\theta = \theta s - \theta e = 56\text{-}20 = 36$ C.^{0}

The energy received by the water over a period of one hour is therefore :

$$E=50 \times 36 \times 5 = 9 .10^6 \text{ J.}$$

The power supplied by the panel is P=E/3600 (since 1 W corresponds to 1J/s):

$$P_{\text{supplied}} = E/3600 = 9000\text{kJ}/3600 = 2500\text{W} = 2.5 \text{ kW.}$$

3- The solar panel receives a power equal to the product of its surface area and the light power received, i.e. : P received = 3 x 1000 = 3000 W = 3 kW.

4-Efficiency is defined as the ratio of power supplied to power received.

$$\eta = 2.5\text{kW} / 3\text{kW} = 0.83 = 83\%. \text{ Therefore } \eta = 83\%.$$

Exercise 7:

Which of these statements are true?

a) Worldwide photovoltaic solar power production capacity has been stable for 10 years.

b) Photovoltaic solar panels are not recyclable

c) Concentrated solar thermal power plants generate electricity

d) A solar thermal panel produces heat

Solutions :

c,d .

Exercise 8:

A solar panel has a peak power of 100 W when it receives a light power PL = 1000 W/m^2 . It is made up of photovoltaic cells connected both in series and in parallel. The cells in each string are connected in series, and the individual strings are connected in parallel. The panel terminal voltage is 40V, and each cell delivers a voltage of 0.5V and a current of 0.5A.

1- How many cells are there in a branch?

2- What is the current flowing through the panel? Deduce the number of panel branches.

3- Determine the total number of cells in the panel.

4- Each cell is a 0.05m square.

a) What is the total surface area of the solar panel?

b) Calculate its energy efficiency η.

Solution:

1-The voltage across a string is the same as that across the panel, since they are connected in parallel $U_B = 40$ V.

In a string, the law of voltage additivity applies, and as it is made up of n cells, we have : $U_B = n.\ U_C$ so $n = U_B / U_C = 40/0.5 = 80$ cells.

A branch consists of n=80 cells.

2-One has P = U.I so I = P/U = 100/40 = 2.5A.

Since the branches are branched, we can apply the law of nodes. Let's note "m" the number of branches, and all of them deliver the same intensity I_1 since they are identical in one branch: $I = m.\ I_1$.

So $m = I/I_1 = 5$. There are m=5 branches.

3- The total number of cells in the panel. N'= n.m =5 x 80= 400 cells.

There are N'=400 cells in this solar panel.

4- Each cell is a 0.05m square.

a-Let "s" **be** the surface area of a cell: $s = a^2 = (5.10\ )^{-22} = 25\ 10^{-4} = 0.25.10^{-2}$ m^2.

The total surface area of the panel is S = N'. s = 400 x $0.25.10^{-2} = 1\ m^2$.

b-Efficiency: $\eta = Pc/\ P_L$ where Pc = 100 W and $P_L = 1000 \times 1 = 1000$ W.

Efficiency: $\eta = Pc/\ P_L = 100$ W /1000 W = 0.10 i.e. $\eta = 10\%$.

Exercise 9:

The characteristics of a photovoltaic module are given in the table below when the module receives a radiant power of 1000 W/m² of module surface.

Caractéristiques électriques (à 1000 W/m²)		
T cellules (°C)	25	50
Pmax (W)	36	32.5
U à Pmax (V)	16.3	14.4
I court-circuit (A)	2.45	2.50
U circuit ouvert (V)	20.3	18.4
I à U = 10(V)	2.29	2.28

1- Plot the voltage-current characteristic (voltage on the x-axis and current on the y-axis) of this photovoltaic module, at 50°C, for a received radiant power of1000 W/m². Place :

a- the operating point A corresponding to the short-circuit current ;

b- operating point B corresponding to an open circuit ;

c- operating point C corresponding to maximum available electrical power.

2- At 50°C, this module receives a radiant power density of 1000 W/m . ²
The voltage at its terminals, when operating, is equal to 10V.

a- According to the data, what is the value of the current I?

b- What is the electrical power supplied?

c- The surface area of the module is 0.185 m². Calculate the module's energy efficiency.

3- What can we conclude about the influence of an increase in temperature on the performance of a photovoltaic solar panel? Is the same true for a solar thermal panel?

Solution:

1-Allure of the voltage-current characteristic of the photovoltaic module.

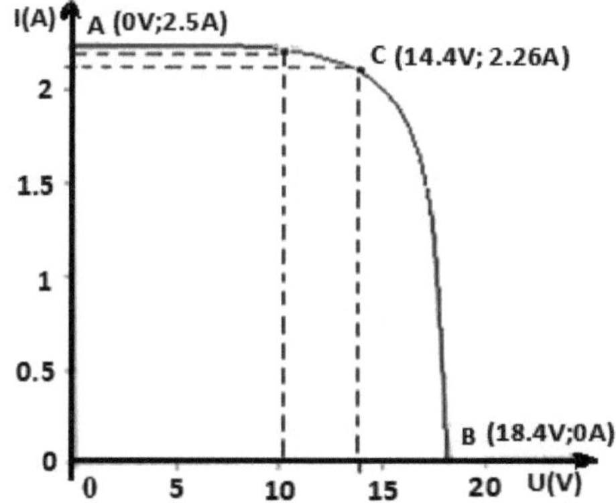

At point A: I cc = 2.5 A. At point B: U co = 18.4 V.

At point C: Pmax = 32.5 W and U opt = 14.4 V => I opt = Pmax / U opt = 2.26 A.

2- At 50°C, this module receives a radiant power density of 1000 W/m^2 .

a) For U = 10V => I = 2.28 A. (according to the curve).

b) $P_{supplied}$ =U.I=10×2.28= 22.8 W and $P_{reçue}$ = 1000(W/m^2) × 0.185(m^2) = 185 W.

c) Efficiency: $\eta = P_{supplied}/P_{reçue}$ = 22.8 W/185W = 12.3%. η =12.3%.

3- Performance is better at 25°C. An increase in temperature therefore reduces the performance of a photovoltaic solar panel. The opposite is true for thermal solar panels.

Exercise 10:

What are the differences between passive and active solar energy?

Solution:

Passive solar heating uses solar energy without the need for special equipment. In fact, it's based on the very design of your home! Passive solar heating is the use of solar energy for heating and cooling. It is based on the architectural design of the house and the choice of materials. Unlike passive solar energy, active solar energy is created by specific equipment that recovers the sun's rays and transforms them into heat (solar thermal), electricity (solar photovoltaic) or heat and electricity (solar thermodynamic).

Printed by Books on Demand GmbH, Norderstedt / Germany